Springer Theses

Recognizing Outstanding Ph.D. Research

Aims and Scope

The series "Springer Theses" brings together a selection of the very best Ph.D. theses from around the world and across the physical sciences. Nominated and endorsed by two recognized specialists, each published volume has been selected for its scientific excellence and the high impact of its contents for the pertinent field of research. For greater accessibility to non-specialists, the published versions include an extended introduction, as well as a foreword by the student's supervisor explaining the special relevance of the work for the field. As a whole, the series will provide a valuable resource both for newcomers to the research fields described, and for other scientists seeking detailed background information on special questions. Finally, it provides an accredited documentation of the valuable contributions made by today's younger generation of scientists.

Theses may be nominated for publication in this series by heads of department at internationally leading universities or institutes and should fulfill all of the following criteria

- They must be written in good English.
- The topic should fall within the confines of Chemistry, Physics, Earth Sciences, Engineering and related interdisciplinary fields such as Materials, Nanoscience, Chemical Engineering, Complex Systems and Biophysics.
- The work reported in the thesis must represent a significant scientific advance.
- If the thesis includes previously published material, permission to reproduce this must be gained from the respective copyright holder (a maximum 30% of the thesis should be a verbatim reproduction from the author's previous publications).
- They must have been examined and passed during the 12 months prior to nomination.
- Each thesis should include a foreword by the supervisor outlining the significance of its content.
- The theses should have a clearly defined structure including an introduction accessible to new PhD students and scientists not expert in the relevant field.

Indexed by zbMATH.

More information about this series at https://link.springer.com/bookseries/8790

Matteo M. Defranchis

First Measurement of the Running of the Top Quark Mass

Doctoral Thesis accepted by
University of Hamburg, Hamburg, Germany

Springer

Author
Dr. Matteo M. Defranchis
Experimental Physics - CMS
CERN
Meyrin, Geneva, Switzerland

Supervisor
Prof. Dr. Katerina Lipka
DESY
Hamburg, Germany

University of Wuppertal
Wuppertal, Germany

ISSN 2190-5053 ISSN 2190-5061 (electronic)
Springer Theses
ISBN 978-3-030-90375-6 ISBN 978-3-030-90376-3 (eBook)
https://doi.org/10.1007/978-3-030-90376-3

This Springer imprint is published by the registered company Springer Nature Switzerland AG
The registered company address is: Gewerbestrasse 11, 6330 Cham, Switzerland

Audentes fortuna iuvat

Latin proverb

Supervisor's Foreword

Mass is one of the most complex concepts in fundamental physics, which went through a long history of conceptual developments. Being first understood in classical mechanics as a measure of inertia, it was later interpreted in the theory of special relativity as a form of energy. The modern quantum field theories combine classical physics, special relativity and quantum mechanics to describe the subatomic world. The most successful of those is the Standard Model of particle physics, which describes the electromagnetic, strong and weak interactions of the elementary particles. The masses of these particles and the strengths of the interactions are fundamental parameters of the Standard Model.

The strong interaction of the elementary constituents of nuclear matter, quarks and gluons, is described by the Quantum Chromodynamics (QCD), which exhibits the property of asymptotic freedom. This property makes it possible to investigate quasi-free quarks at very small distances and leads to the fact that the coupling of strong interaction is not a fixed value, but depends on the energy scale at which it is probed.

According to QCD, the strong coupling constant rapidly decreases with growing energy scales—the effect referred to as "running".

The same is also true for the masses of the quarks, which can themselves be understood as fundamental couplings, for example, in relation to the interaction with the Higgs field. The top quark is not only the most massive quark, it is the most massive elementary particle of all. Its mass, which is about 170 times the mass of the proton, plays a key role in the strong and in the electroweak interactions and determines the stability of the electroweak vacuum. Not only the value and precision of the top quark mass are important but also its behaviour.

In Quantum Chromodynamics, the energy scale dependence of the strong coupling constant and of the quark masses is described by the QCD Renormalization Group Equations. The experimental verification of the running of strong coupling and of quark masses is a fundamental test of the validity of Quantum Chromodynamics. Furthermore, effects of physics beyond the Standard Model can lead to modifications in the running at high energy scales, probed by the Large Hadron Collider at CERN. Therefore, the measurement of this effect is also a search for unknown physics. The running of the strong coupling constant has been established in many experiments in the past decades, on ranges of scales that span over three orders of magnitude. In addition, evidence was found for the running of the masses of the charm and beauty quarks.

In his thesis, Matteo M. Defranchis investigates for the first time the running of the top quark mass, using the production of top quark-antiquark pairs in proton-proton collisions at the LHC at a centre-of-mass energy of 13 TeV, recorded by the Compact Muon Solenoid (CMS) experiment.

Experimentally, interesting events are selected by searching for the specific decay products of a top quark-antiquark pair. In the vast majority of cases, top quarks decay into an energetic jet and a W boson, which in turn can decay into a lepton and a neutrino. Jets and leptons can be identified and measured with high precision by the CMS detector, while neutrinos escape undetected and reveal themselves as missing energy.

In order to probe the running of the top quark mass, the production rate of the top quark-antiquark pairs is measured at different energy scales. From this measurement, the top quark mass is extracted at those energy scales using theory predictions implying so-called minimal-subtraction renormalization scheme.

The pioneering result of this thesis demonstrates the fundamental quantum effect of the running of the quark masses for the most massive elementary particle, the top quark, probing the energy scales up to 1 TeV, and is a big step forward in the field of particle physics.

June 2021

Katerina Lipka
DESY
Hamburg, Germany

University of Wuppertal
Wuppertal, Germany

Abstract

In this thesis, the first experimental determination of the running of the top quark mass is presented. The running is extracted from a measurement of the differential top quark-antiquark ($\mathrm{t\bar{t}}$) production cross section as a function of the invariant mass of the $\mathrm{t\bar{t}}$ system, $m_{\mathrm{t\bar{t}}}$. The analysis is performed using proton-proton collision data recorded by the CMS detector at the CERN LHC in 2016, corresponding to an integrated luminosity of 35.9 fb^{-1}. Candidate $\mathrm{t\bar{t}}$ events are selected in the final state with an electron and a muon of opposite charge, and the differential cross section $\mathrm{d}\sigma_{\mathrm{t\bar{t}}}/\mathrm{d}m_{\mathrm{t\bar{t}}}$ is determined at the parton level by means of a maximum-likelihood fit to multidifferential final-state distributions. The value of the top quark mass in the modified minimal subtraction ($\overline{\mathrm{MS}}$) renormalization scheme, $m_{\mathrm{t}}(\mu)$, is determined as a function of the scale $\mu = m_{\mathrm{t\bar{t}}}$ by comparing the measured $\mathrm{d}\sigma_{\mathrm{t\bar{t}}}/\mathrm{d}m_{\mathrm{t\bar{t}}}$ to theoretical predictions at next-to-leading order, and the resulting scale dependence is interpreted as the running of the top quark mass. The observed running is found to be compatible with the one-loop solution of the corresponding renormalization group equation, up to a scale of the order of 1 TeV.

Publications related to this thesis

- CMS Collaboration, Running of the top quark mass from proton-proton collisions at $\sqrt{s} = 13\,\text{TeV}$, *Phys. Lett. B* **803** (2020) 135263 [arXiv:1909.09193].
- CMS Collaboration, Measurement of the $t\bar{t}$ production cross section, the top quark mass, and the strong coupling constant using dilepton events in pp collisions at $\sqrt{s} = 13\,\text{TeV}$, *Eur. Phys. J. C* **79** (2019) 368 [arXiv:1812.10505].
- CMS Collaboration, Identification of heavy-flavour jets with the CMS detector in pp collisions at 13 TeV, *JINST* **13** (2018) P05011 [arXiv:1712.07158].
- CMS Collaboration, B-jet trigger performance in Run 2, CMS Detector Performance Note CMS-DP-2019-042 [http://cds.cern.ch/record/2708546].

Acknowledgements

I would like to express my gratitude to everyone at DESY and in CMS who contributed, directly or indirectly, to the outcome of this work and to the development my personal and scientific maturity during the years of my Ph.D. The biggest thanks of all goes to my supervisor Katerina Lipka. Her wise and enthusiastic supervision has been a guidance at all stages and in all aspects of my Ph.D. I am also grateful to her for giving me the opportunity to develop my skills and career with a large degree of independence, and for her tireless personal and professional support. I am also grateful to Sven Moch, my second supervisor, for the useful theoretical discussions and his essential contribution to this work, and to Jan Kieseler for his invaluable support to this project. I would also like to thank Olaf Behnke for the vital help with the statistical aspects of this work, Juan Fuster for the interesting discussions on top quark mass measurements and for his essential support to the development of my career, and Marcel Vos for reviewing this thesis.

I am also grateful to all the members of the DESY Top group who helped creating a stimulating and friendly working environment, and in particular to Maria Aldaya and Nadjieh Jafari for their wise and passionate leadership. I would also like to thank Mykola Savitskyi and Till Arndt for their help at the beginning of my Ph.D., Petra Van Mulders, who greatly contributed to the review of the differential measurement, Martijn Mulders, who closely followed all the analyses presented in this thesis, and everyone in the DESY and CMS Top groups who contributed to the review of these results. I am also grateful to all the people in the b-tagging group, in particular Ivan Marchesini, Kirill Skovpen and Mauro Verzetti, who gave me the great opportunity to take responsibility of the high-level trigger subgroup. I would also like to thank my co-convener Luca Mastrolorenzo, the outgoing and incoming conveners Xavier Coubez and John Alison and everyone who contributed to the activities of the group, in particular Marina Kolosova.

Finally, I would like to thank all my friends, old and new, my colleagues and office mates and everyone who helped making these years special. Their friendship, company and support have been invaluable. I would also like to remember my friend Jihyun, knowing her has been an honour to me.

Matteo M. Defranchis

Contents

Chapter 1
Introduction

The standard model (SM) of particle physics is a quantum field theory that successfully describes the interactions between elementary particles. The interactions are defined by the principle of local gauge invariance, which requires that the Lagrangian density describing the dynamics of the fields is invariant under local gauge transformations. The SM comprises an electroweak sector, based on the $SU(2)_L \otimes U(1)_Y$ symmetry group, and a strong sector, described by quantum chromodynamics (QCD) and based on the $SU(3)_c$ group. The electroweak symmetry is spontaneously broken by the Higgs mechanism, which is responsible for generating the masses of the electroweak bosons and the fundamental fermions. The validity of the SM has been tested up to energy scales of the order of a few TeV, and the discovery of the Higgs boson by the ATLAS and CMS experiments at the CERN LHC in 2012 [1, 2] confirmed the existence of the remaining undiscovered particle predicted by the SM.

In the standard model, a special role is played by the top quark, discovered in 1995 by the CDF and D0 experiments at the Fermilab Tevatron [3, 4]. With a mass of about 173 GeV [5], the top quark is the most massive elementary particle currently known. For this reason, unlike other quarks, top quarks decay before forming bound states. This provides a unique opportunity to study the properties of an unconfined quark, such as its spin and charge. Furthermore, due to its large Yukawa coupling, quantum loops containing top quarks are responsible for large corrections to the Higgs mass and self coupling. The size of these corrections is proportional to the value of the top quark mass, m_{t}, which makes the precise knowledge of this parameter an essential ingredient for understanding the mechanism of electroweak symmetry breaking.

However, beyond leading order in perturbation theory, quantum corrections to the quarks and gluon propagators and to the QCD vertices lead to ultraviolet divergences in the definition of the parameters of the QCD Lagrangian, i.e. the strong coupling constant (α_{S}) and the quark masses. With a procedure known as renormalization, these infinities can be reabsorbed in the definition of the bare (i.e. tree-level) parameters. However, the definition of the QCD parameters depends on the renor-

M. M. Defranchis, *First Measurement of the Running of the Top Quark Mass*,
Springer Theses, https://doi.org/10.1007/978-3-030-90376-3_1

malization procedure (or scheme), which is not unique. In the modified minimal subtraction ($\overline{\text{MS}}$) scheme, α_S and the quark masses depend on the scale at which the infinities are subtracted, known as the renormalization scale. The resulting scale dependence, or running, is described by a set of renormalization group equations (RGEs). The solutions to the RGEs, which can be obtained in perturbation theory, predict that the values of α_S and the quark masses decrease with increasing energy scales. This fact is also known as the asymptotic freedom of QCD, and ensures the validity of the perturbative approach at sufficiently high energies. The experimental determination of the running of α_S and of the quark masses is a powerful probe of the validity of perturbative QCD, but also an indirect search for new phenomena. In fact, the RGEs can be modified by contributions of physics beyond the standard model, e.g. in the context of supersymmetric theories [6] or in scenarios where the fermion masses are generated dynamically [7]. The running of α_S has been verified on a range of energies covering several orders of magnitude using data from different experiments [5]. Studies on the running of the bottom quark mass have also been performed [8] using data from various experiments at the CERN LEP [9–12], SLAC SLC [13], and DESY HERA [14] colliders, and the running of the charm quark mass has been probed using measurements of charm quark production in deep inelastic scattering at the DESY HERA [15]. The running of the top quark mass, instead, is investigated for the first time in this thesis.

Another procedure used to renormalise the quark masses is represented by the on-shell (or pole) scheme, in which the mass parameters are redefined order-by-order in perturbation theory so that they correspond to the poles of the quark propagators. However, it is known that the quark pole masses are affected by a fundamental ambiguity of the order of $\Lambda_{\text{QCD}} \simeq 250\,\text{MeV}$ [16–18].

A common way to measure the mass of the top quark is to reconstruct the invariant mass of its decay products, or a related observable. However, these measurements heavily rely on the shape of the final-state distributions obtained using Monte Carlo (MC) simulation. Therefore, the measured value of m_t lacks a clear theoretical interpretation, and is understood as the value of the top quark mass parameter in the simulation (m_t^{MC}) that describes the data best. Alternatively, the value of m_t can be determined in a well-defined renormalization scheme by comparing fixed-order theoretical calculations of observables sensitive to the value of m_t to the corresponding measurements. At a hadron collider, where top quarks are mainly produced as quark-antiquark ($t\bar{t}$) pairs, these observables are usually the inclusive and differential $t\bar{t}$ production cross sections. However, measurements of these observables are affected by a residual dependence on the value of m_t^{MC}, which has to be taken into account when extracting m_t. This is usually done by assuming a certain relation between m_t and m_t^{MC}, which introduces an additional uncertainty.

In this thesis, the value of $m_t(m_t)$, i.e. the top quark mass in the $\overline{\text{MS}}$ scheme evaluated at the scale $\mu = m_t$, is determined by comparing a measurement of the inclusive $t\bar{t}$ production cross section to next-to-next-to-leading order (NNLO) theoretical predictions. The same method is also used to extract the value of α_S at the mass of the Z boson, $\alpha_S(m_Z)$, and the top quark pole mass, m_t^{pole}. These results are obtained using proton-proton collision data at a centre-of-mass energy of 13 TeV collected by the CMS experiment at the CERN LHC in 2016, corresponding to an integrated luminosity of 35.9 fb^{-1}. Candidate $\mathrm{t\bar{t}}$ events are selected in the final state with an electron and a muon of opposite charge, and the inclusive $\mathrm{t\bar{t}}$ cross section $\sigma_{\mathrm{t\bar{t}}}$ is measured simultaneously with $m_{\mathrm{t}}^{\mathrm{MC}}$ by means of a maximum-likelihood fit to multidifferential final-state distributions. With this method, the experimental dependence of the measured $\sigma_{\mathrm{t\bar{t}}}$ on the value of $m_{\mathrm{t}}^{\mathrm{MC}}$ is mitigated, and no prior assumption on the relation between $m_{\mathrm{t}}^{\mathrm{MC}}$ and m_{t} has to be made in the extraction of m_{t} [19].

The method used to measure $\sigma_{\mathrm{t\bar{t}}}$ is then extended to determine the differential $\mathrm{t\bar{t}}$ cross section as a function of the invariant mass of the $\mathrm{t\bar{t}}$ system, $m_{\mathrm{t\bar{t}}}$. The differential cross section $\mathrm{d}\sigma_{\mathrm{t\bar{t}}}/\mathrm{d}m_{\mathrm{t\bar{t}}}$ is determined directly at the parton level by embedding the response matrix in the likelihood function. The value of $m_{\mathrm{t}}(\mu)$ is then extracted as a function of the scale $\mu = m_{\mathrm{t\bar{t}}}$ by comparing theoretical predictions at next-to-leading order (NLO) to the measured $\mathrm{d}\sigma_{\mathrm{t\bar{t}}}/\mathrm{d}m_{\mathrm{t\bar{t}}}$. The resulting scale dependence is interpreted as the running of the top quark mass in the $\overline{\mathrm{MS}}$ scheme, which is probed up to a scale of the order of 1 TeV.

This thesis is organized as follows. The theoretical aspects of highest relevance for this work are illustrated in Chap. 2. A brief description of the CERN LHC and the CMS detector is provided in Chap. 3, and the most relevant aspects of the CMS event reconstruction algorithms are illustrated in Chap. 4. The same chapter also contains studies related to the identification of b jets in boosted topologies and at the CMS high-level trigger obtained with significant contribution from the author of this thesis, and published in Refs. [20] and [21], respectively. The simultaneous measurement of $\sigma_{\mathrm{t\bar{t}}}$ and $m_{\mathrm{t}}^{\mathrm{MC}}$ is then described in detail in Chap. 5, and the extraction of α_S and m_{t} is discussed in Chap. 6. The results presented in the latter two chapters, obtained for the purpose of this thesis, are also documented in Ref. [22]. Finally, the first measurement of the running of the top quark mass, entirely developed in the scope of this thesis, is presented in Chap. 7 and documented in Ref. [23].

References

1. *ATLAS* Collaboration, Observation of a new particle in the search for the standard model Higgs boson with the ATLAS detector at the LHC. Phys Lett B 716:1 (2012). https://doi.org/10.1016/j.physletb.2012.08.020, arXiv:1207.7214
2. *CMS* (2012) Collaboration, Observation of a new boson at a mass of 125 GeV with the CMS experiment at the LHC. Phys Lett B 716:30. https://doi.org/10.1016/j.physletb.2012.08.021, arXiv:1207.7235
3. *CDF* (1995) Collaboration, Observation of top quark production in $\bar{\mathrm{p}}\mathrm{p}$ collisions. Phys. Rev. Lett. 74:2626. https://doi.org/10.1103/PhysRevLett.74.2626, arXiv:hep-ex/9503002
4. *D0* (1995) Collaboration, Observation of the top quark. Phys Rev Lett 74:2632 https://doi.org/10.1103/PhysRevLett.74.2632, arXiv:hep-ex/9503003
5. Particle Data Group, Tanabashi M et al (2018) Review of particle physics. Phys Rev D 98:030001. https://doi.org/10.1103/PhysRevD.98.030001
6. Mihaila L (2013) Precision calculations in supersymmetric theories. Adv High Energy Phys 2013:607807. https://doi.org/10.1155/2013/607807, arXiv:1310.6178

7. Christensen ND, Shrock R (2005) Implications of dynamical generation of standard-model fermion masses. Phys Rev Lett 94:241801. https://doi.org/10.1103/PhysRevLett.94.241801, arXiv:hep-ph/0501294
8. Behnke O, Geiser A, Lisovyi M (2015) Charm, beauty and top at HERA. Prog Part Nucl Phys 84:1. https://doi.org/10.1016/j.ppnp.2015.06.002, arXiv:1506.07519
9. *DELPHI* (2008) Collaboration, study of b-quark mass effects in multijet topologies with the DELPHI detector at LEP. Eur Phys J C 55:525. https://doi.org/10.1140/epjc/s10052-008-0631-5, arXiv:0804.3883
10. *DELPHI* (2006) Collaboration, Determination of the b quark mass at the M_Z scale with the DELPHI detector at LEP. Eur Phys J C 46:569. https://doi.org/10.1140/epjc/s2006-02497-6
11. *ALEPH* (2000) Collaboration, A Measurement of the b quark mass from hadronic Z decays. Eur Phys J C **18**:1 (2000). https://doi.org/10.1007/s100520000533, arXiv:hep-ex/0008013
12. *OPAL* (2001) Collaboration, Determination of the b quark mass at the Z mass scale. Eur Phys J C 21:411. https://doi.org/10.1007/s100520100746, arXiv:hep-ex/0105046
13. Brandenburg A, Burrows PN, Muller D, Oishi N, Uwer P (1999) Measurement of the running b quark mass using $e^+e^- \to b\bar{b}g$ events. Phys Lett B 468:168 (1999). https://doi.org/10.1016/S0370-2693(99)01194-6, arXiv:hep-ph/9905495
14. *ZEUS* (2014) Collaboration, Measurement of beauty and charm production in deep inelastic scattering at HERA and measurement of the beauty-quark mass. JHEP 09:127. https://doi.org/10.1007/JHEP09(2014)127, arXiv:1405.6915
15. Gizhko A et al (2017) Running of the charm-quark mass from HERA deep-inelastic scattering data. Phys Lett B 775:233. https://doi.org/10.1016/j.physletb.2017.11.002, arXiv:1705.08863
16. Beneke M, Braun VM (1994) Heavy quark effective theory beyond perturbation theory: renormalons, the pole mass and the residual mass term. Nucl Phys B 426:301. https://doi.org/10.1016/0550-3213(94)90314-X, arXiv:hep-ph/9402364
17. Bigi IIY, Shifman MA, Uraltsev NG, Vainshtein AI (1994) The Pole mass of the heavy quark. Perturbation theory and beyond. Phys Rev D 50:2234 https://doi.org/10.1103/PhysRevD.50.2234, arXiv:hep-ph/9402360
18. Smith MC, Willenbrock SS (1997) Top quark pole mass. Phys Rev Lett 79:3825. https://doi.org/10.1103/PhysRevLett.79.3825, arXiv:hep-ph/9612329
19. Kieseler J, Lipka K, Moch S (2016) Calibration of the top-quark monte carlo mass. Phys Rev Lett 116:162001. https://doi.org/10.1103/PhysRevLett.116.162001, arXiv:1511.00841
20. *CMS* Collaboration, Identification of heavy-flavour jets with the CMS detector in pp collisions at 13 TeV. JINST 13, 05011. https://doi.org/10.1088/1748-0221/13/05/P05011, arXiv:1712.07158
21. *CMS* (2019) Collaboration, B-jet trigger performance in Run 2, CMS Detector Performance Note CMS-DP-2019-042, CERN, Geneva, Nov 2019. http://cds.cern.ch/record/2708546
22. *CMS* (2019) Collaboration, Measurement of the $t\bar{t}$ production cross section, the top quark mass, and the strong coupling constant using dilepton events in pp collisions at $\sqrt{s} = 13$TeV. Eur Phys J C 79:368. https://doi.org/10.1140/epjc/s10052-019-6863-8, arXiv:1812.10505
23. *CMS* (2020) Collaboration, Running of the top quark mass from proton-proton collisions at $\sqrt{s} = 13$TeV. Phys Lett B 803:135263. https://doi.org/10.1016/j.physletb.2020.135263, arXiv:1909.09193

Chapter 2
The Standard Model of Particle Physics and the Top Quark

This Chapter provides an overview of the theoretical aspects underlying the work described in this thesis, and is structured as follows: a brief introduction to the standard model of particle physics is given in Sect. 2.1. Details about the electroweak sector and the Higgs mechanism are discussed in Sects. 2.2 and 2.3, respectively. The theory of quantum chromodynamics is then described in Sect. 2.4, with particular emphasis given to the phenomenological implications of renormalization. The factorization properties of QCD are then discussed in Sect. 2.5, and selected aspects of Monte Carlo simulation are illustrated in Sect. 2.6. After this general introduction, the properties of the top quark and its phenomenology at the Large Hadron Collider (LHC) are illustrated in Sect. 2.7, with particular emphasis given to the aspects relevant for this thesis. Finally, details on the fixed-order calculations and the MC simulation used in this work are given in Sects. 2.8 and 2.9, respectively.

2.1 The Standard Model of Particle Physics

The standard model (SM) of particle physics is a quantum field theory that describes the fundamental interactions between elementary particles. The interactions are defined by the paradigm of local gauge invariance, which requires that the Lagrangian density describing the dynamics of the fields is invariant under local gauge transformations. The SM is divided in a unified electroweak sector, based on the $SU(2)_L \otimes U(1)_Y$ symmetry group, and a strong sector, based on the $SU(3)_c$ group.

The fundamental particles of the SM and their interactions are illustrated in Fig. 2.1. Quarks and leptons are the fundamental building blocks of matter, and are described by fermionic fields of spin $1/2$ obeying the Dirac equation. The strong and electroweak interactions between fermions are mediated by bosons of spin 1, corresponding to the gauge fields of the respective symmetry group. In particular, the

M. M. Defranchis, *First Measurement of the Running of the Top Quark Mass*,
Springer Theses, https://doi.org/10.1007/978-3-030-90376-3_2

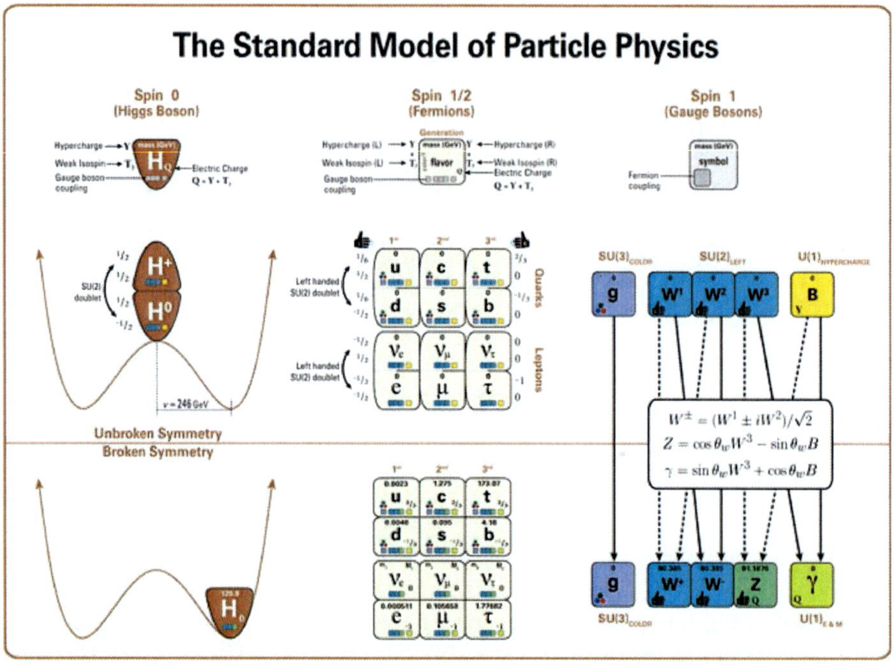

Fig. 2.1 The fundamental particles of the standard model (quarks and leptons) and the bosons responsible for their interactions [1]

electromagnetic interaction is mediated by the photon (γ), the charged- and neutral-current weak interactions are mediated by the $W^{\pm}$ and Z bosons, respectively, and the strong interaction is mediated by the gluon (g). A different kind of interaction is mediated by the Higgs boson (H), a scalar field which is responsible for generating the mass of all elementary particles through the mechanism of spontaneous symmetry breaking, as illustrated in Fig. 2.1.

Leptons are classified as charged or neutral leptons. The charged leptons are the electron (e), the muon (μ) and the τ lepton. Each charged lepton is associated to a corresponding neutral lepton, called neutrino, which therefore appear in three different flavours: electron neutrinos (ν_e), muon neutrinos (ν_μ) and τ neutrinos (ν_τ). The charged-current weak interaction couples each charged lepton to the corresponding neutrino, and only the left-handed components of the fields participate in the interaction [2, 3].

Quarks are divided into up-type and down-type quarks, with electric charge $+2/3$ and $-1/3$, respectively, in units of the electron charge. With the exception of the top quark, discussed in detail in Sect. 2.7, quarks are always observed in bound states called hadrons. The up (u), down (d), and strange (s) quarks are commonly referred to as light quarks, while the charm (c), bottom (b), and top (t) quarks are known as heavy quarks. Quarks participate in all known fundamental interactions. The charged-current weak interaction couples up-type quarks to down-type quarks and, unlike in the lepton sector, the interaction is not diagonal in the flavour. This property of the quark sector is known as flavour mixing.

In the SM, quarks and leptons can be classified into three generations, corresponding to the columns in Fig. 2.1. Each generation has identical physical properties, with the exception of the masses of the particles. In particular, the first generation contains the lightest charged lepton and its neutrino and the lightest up-type and down-type quarks, while the third generation contains the heaviest ones. The number of neutrinos (and therefore of lepton generations) has been tested with high precision by the experiments at the CERN LEP collider, and was found to be compatible with three [4].

2.2 The Unified Theory of Electroweak Interactions

Electromagnetic and weak interactions are successfully described by the Glashow-Weinberg-Salam (GWS) theory [5–7], which is a spontaneously broken gauge theory based on the non-Abelian symmetry group $SU(2)_L \otimes U(1)_Y$. The left-handed components of the fermionic fields are organized in isospin doublets of $SU(2)_L$:

$$L_L^j = \begin{pmatrix} \nu_{\ell L}^j \\ \ell_L^j \end{pmatrix}, \quad Q_L^{\prime j} = \begin{pmatrix} q_{uL}^{\prime j} \\ q_{dL}^{\prime j} \end{pmatrix}, \quad j = 1, 2, 3,$$

where ν_ℓ^j, ℓ^j, $q_u^{\prime j}$, and $q_d^{\prime j}$ indicate the Dirac fields of the lepton neutrinos, charged leptons, up-type quarks, and down-type quarks, respectively, and the index j runs over the fermion generations. The right-handed components of the fields, instead, are treated as $SU(2)_L$ singlets.

For each generation, the gauge-invariant Lagrangian density describing the dynamics of fermion fields can be written as:

$$\mathcal{L} = i\overline{L}_L \gamma^\mu D_\mu L_L + i\overline{\nu}_{\ell R} \gamma^\mu D_\mu \nu_{\ell R} + i\overline{\ell}_R \gamma^\mu D_\mu \ell_R, \tag{2.1}$$

$$D_\mu = \partial_\mu - igT_i W_\mu^i(x) - ig' \frac{Y_\psi}{2} B_\mu(x), \tag{2.2}$$

where g, g' are real dimensionless parameters, Y_ψ is the hypercharge operator, T_i are the three generators of $SU(2)_L$, γ^μ are the four Dirac matrices, and $W_\mu^i(x)$ and $B_\mu(x)$ are real fields. When acting on an isospin doublet, the T_i correspond to $\sigma_i/2$, where σ_i are the three Pauli matrices. The experimentally observed charged-current interactions are derived by performing a linear combinations of the W_μ^1 and W_μ^2 fields:

$$W_\mu^\pm = \frac{1}{\sqrt{2}} \left(W_\mu^1 \mp W_\mu^2 \right), \tag{2.3}$$

while the electromagnetic field A_μ can be obtained by performing a rotation of the remaining fields:

$$\begin{pmatrix} B_\mu \\ W_\mu^3 \end{pmatrix} = \begin{pmatrix} \cos\theta_W & -\sin\theta_W \\ \sin\theta_W & \cos\theta_W \end{pmatrix} \begin{pmatrix} A_\mu \\ Z_\mu \end{pmatrix}, \tag{2.4}$$

under the condition $g\sin\theta_W = g'\cos\theta_W = e$, where e is the electric charge of the electron. The parameter θ_W is referred to as the electroweak mixing angle, or Weinberg angle, and the neutral field Z_μ corresponds to the field of the Z boson. The effective value of $\sin^2\theta_W$ was measured with high precision at the Z pole by the experiments at the CERN LEP, and was found to be 0.23153 ± 0.00016 [4].

In this formalism, the relation between the weak isospin, the electric charge, and the hypercharge of a fundamental particle is given by the Gell-Mann-Nishijima formula:

$$Q = T_3 + \frac{Y_\psi}{2}, \tag{2.5}$$

where Q is the electric charge operator. It follows from Eq. 2.5 that the hypercharge of right-handed neutrinos must be zero, which implies that they do not participate in any SM interaction. For this reason, the corresponding fields are removed from the Lagrangian density in Eq. 2.1.

The validity of the GWS model has been extensively tested with a high level of precision, e.g. by the experiments at the CERN LEP [4], and remarkable agreement between the theoretical predictions and the measurements has been observed.

2.3 The Higgs Mechanism and Quark Mixing

The formalism illustrated in Sect. 2.2 does not allow mass terms for fermions and electroweak bosons to be introduced without explicitly breaking the gauge symmetry. The problem is solved by the Higgs-Brout-Englert mechanism [8, 9], which generates the desired mass terms through the mechanism of spontaneous symmetry breaking. An isospin doublet of scalar fields ϕ is introduced, which is assumed to be subject to the potential:

$$V(\phi) = -\mu^2\,|\phi|^2 + \lambda\,|\phi|^4, \tag{2.6}$$

where μ^2 and λ are real parameters. This potential is invariant under a $SU(2)_L \otimes U(1)_Y$ gauge transformation, and can therefore be included in the GSW Lagrangian. In order to ensure the stability of the electroweak vacuum, the parameter λ is required to be positive. Moreover, for positive values of μ^2, the ground states of the potential corresponds to an infinite number of degenerate fields satisfying the condition $|\phi|^2 = \mu^2/(2\lambda) \equiv v^2/2$. In this case, the ground-state fields are not invariant under the $SU(2)_L \otimes U(1)_Y$ gauge transformation and develop a non-zero vacuum expectation value corresponding to v. The shape of the Higgs potential for $\mu^2 > 0$ and $\lambda > 0$ is shown in Fig. 2.1, where the physical meaning of the parameter v is also illustrated.

The ground-state field can be expanded around the minimum of the potential and parametrized in terms of its four degrees of freedom:

$$\phi = \frac{1}{\sqrt{2}} e^{i \frac{\sigma_j \theta^j(x)}{v}} \begin{pmatrix} 0 \\ v + H(x) \end{pmatrix}, \tag{2.7}$$

where $H(x)$ and the three θ_j are real scalar fields. It is possible to perform a gauge transformation of the field ϕ so that the phase in Eq. 2.7 vanishes: this gauge is commonly referred to as the unitary gauge. With this choice, which spontaneously breaks the $SU(2)_L \otimes U(1)_Y$ symmetry, the three degrees of freedom carried by the fields θ^j are converted into mass terms for the W and Z boson fields, with $m_\mathrm{W} = gv/2$ and $m_\mathrm{Z} = m_\mathrm{W}/\cos\theta_W$. The obtained mass terms are proportional to $v \simeq 246\,\mathrm{GeV}$ [10], which can be considered the natural scale of the electroweak theory. The field H, which is interpreted as the Higgs boson field, also acquires a mass $m_\mathrm{H} = \sqrt{2\lambda}\, v$, and interaction terms between the Higgs boson and massive vector bosons are naturally generated. Furthermore, the expansion of the potential around the minimum of ϕ gives rise to trilinear and quartic self-interaction terms for the field H, with μ^2/v and λ being the respective couplings.

Fermion masses are obtained by introducing a gauge-invariant Yukawa-like interaction term, that in the case of quarks can be written as:

$$\mathcal{L}_Y^q = -\overline{Q}'^i_L \phi \Gamma'^{ij}_d q'^j_{dR} - \overline{Q}'^i_L \phi_c \Gamma'^{ij}_u q'^j_{uR} + h.c. \tag{2.8}$$

Here, Γ_u and Γ_d are complex matrices in the space of the three generations, $h.c.$ denotes the hermitian conjugate, and $\phi_c = i\sigma_2\phi^*$. It is possible to find a bi-unitary transformation for which $\Gamma_u^{ij} = V_L^{u\dagger}\Gamma'^{ij}_u V_R^u$ and $\Gamma_d^{ij} = V_L^{d\dagger}\Gamma'^{ij}_d V_R^d$ are diagonal matrices with positive real elements, which are interpreted as the quark masses. Flavour-diagonal interaction terms between the Higgs and quark fields are also generated, with coupling constant $y_q = \sqrt{2} m_q/v$, where m_q is the quark mass. The quantity y_q is referred to as Yukawa coupling, and the physical fields are identified with those that diagonalize the mass matrices.

The Higgs boson was discovered in 2012 [11, 12] by the ATLAS and CMS Collaborations at the CERN LHC, and its mass is currently measured to be $125.10 \pm 0.14\,\mathrm{GeV}$ [10]. The properties of the Higgs boson, and in particular the Yukawa couplings of quarks and leptons, are currently being investigated at the LHC. So far, remarkable agreement between the SM predictions and the measurements is observed [13, 14]. Furthermore, the ATLAS and CMS Collaborations recently observed the associated production of a top quark pair and a Higgs boson [15, 16], and the decays of the Higgs boson into b quarks [17, 18] and τ leptons [19, 20]. These results represent a direct evidence of the coupling of the Higgs boson to third-generation fermions. However, only upper bounds are currently available for the Yukawa couplings to second-generation fermions [21–24], while those to first-generation fermions are still out of reach.

The Higgs mechanism can also provide an explanation for the flavour mixing and CP violation observed in the SM [10]. In fact, the charged-current weak interaction can be expressed in terms of the physical quark fields as:

$$J_q^\mu = \sum_{i,j} \overline{q}^i_{uL} \gamma^\mu V_{ij} q^j_{dL}, \tag{2.9}$$

where $V_{ij} = V_L^{u\dagger} V_L^d$ is the Cabibbo-Kobayashi-Maskawa (CKM) matrix [25, 26]. In the SM, the CKM matrix is required to be unitary, and can be expressed in terms of three mixing angles and one CP-violating phase. The elements of the CKM matrix are being measured with increasing precision [10], and tests of unitarity are being performed using data from different experiments [27].

The flavour mixing in the leptonic sector can be described in an analogous way. However, since right handed neutrino fields do not appear in the GWS Lagrangian, it is possible to perform a transformation of the fields without introducing lepton-flavour mixing. Therefore, unlike in the quark sector, the charged-current interactions between leptons are diagonal in the flavour, and mass terms are generated only for charged leptons. However, recent experimental observations of neutrino oscillations indicate that neutrinos are massive particles [28–30]. It is possible to extend the SM Lagrangian in order to obtain mass terms for neutrinos, which leads to the introduction of the counterpart of the CKM matrix in the leptonic sector, known as the Pontecorvo-Maki-Nakagawa-Sakata (PMNS) matrix [31, 32]. The PMNS matrix can be expressed in terms of three mixing angles and one CP-violating phase if neutrinos are Dirac particles, while it requires two additional weak phases if neutrinos are Majorana particles [33]. The elements of the PMNS matrix can be determined via global fit to data of various experiments [34], while the Dirac or Majorana nature of neutrinos is still being investigated [35]. The values of neutrino masses are currently unknown, and only upper limits have been determined [36–39].

2.4 Quantum Chromodynamics and Renormalization

The strong interaction between quarks and gluons is successfully described by quantum chromodynamics (QCD), a gauge theory based on the $SU(3)_c$ symmetry group. The quantum number responsible for strong interactions is the colour charge, which was initially introduced to explain the observed total hadronic cross section in e^+e^- collisions [10] and the existence of the Δ^{++} baryon [40–42]. Quark fields are organized in colour triplets $\psi = (q_a, q_b, q_c)^{\mathrm{T}}$ of the $SU(3)_c$ group, and their dynamics is described by the Lagrangian density:

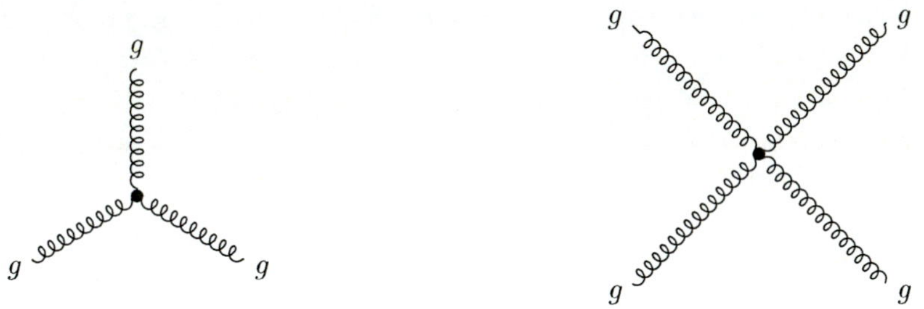

Fig. 2.2 Feynman diagrams corresponding to the trilinear (left) and quartic (right) gluon self-interaction vertices

$$\mathcal{L}_{\text{QCD}} = -\frac{1}{4}F^a_{\mu\nu}F^{\mu\nu}_a + \sum_f \overline{\psi}_f(i\gamma^\mu D_\mu - m_f)\psi_f, \tag{2.10}$$

$$D_\mu = \partial_\mu - ig_s t_a A^a_\mu(x), \tag{2.11}$$

$$F^a_{\mu\nu} = \partial_\mu A^a_\nu - \partial_\nu A^a_\mu + g_s f^{abc} A^b_\mu A^c_\nu. \tag{2.12}$$

Here, f is an index that runs over the quark flavours, g_s is a dimensionless real parameter, t^a are the eight generators of the fundamental representation of $SU(3)_c$, $A^a_\mu(x)$ are the eight gauge fields corresponding to the gluon fields, and f^{abc} are the completely antisymmetric $SU(3)_c$ structure constants defined by the commutation relation $[t_a, t_b] = i f^{abc} t_c$. Gluons couple to quarks with the same flavour and different colours, and are therefore coloured force carriers. Due to the non-Abelian structure of $SU(3)_c$, the dynamic term of the gluon fields generates the trilinear and quartic gluon self-interaction vertices shown in Fig. 2.2. In analogy to quantum electrodynamics, a strong coupling constant is defined as $\alpha_S = g_s^2/(4\pi)$. QCD was experimentally established as a gauge theory with the discovery of the gluon at the DESY PETRA storage ring [43].

Beyond leading order in perturbation theory, corrections to the gluon and quark propagators, as the ones shown in Fig. 2.3, lead to logarithmic ultraviolet divergences in the definition of the parameters of the QCD Lagrangian, i.e. α_S and the quark masses. In local gauge theories, at every order in perturbation theory, infinities can be subtracted and reabsorbed in the definition of the bare (i.e. tree-level) parameters of the Lagrangian [44], which are regarded as unphysical quantities. This process is know as renormalization. The choice of the renormalization scheme, and therefore the definition of the parameters, is not unique. However, at all orders in perturbation theory, the results of calculations of physical observables are independent of the renormalization scheme used.

The modified minimal subtraction ($\overline{\text{MS}}$) scheme is one of the most widely used renormalization procedures. In this scheme, the subtraction of ultraviolet divergences

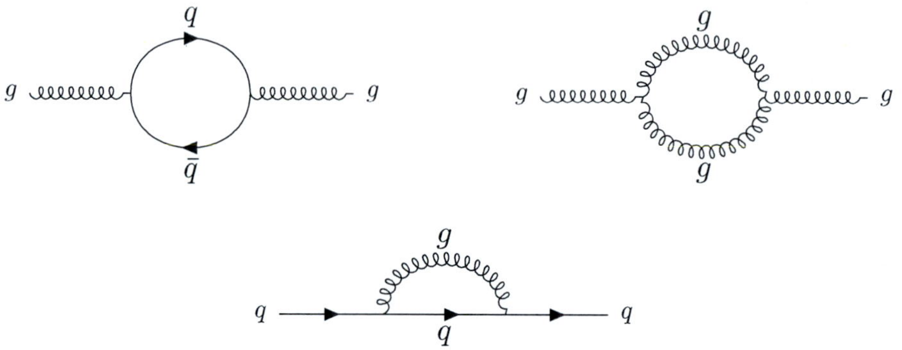

Fig. 2.3 One-loop corrections to the gluon (upper) and quark (lower) propagators which lead to the renormalization of the strong coupling constant and the quark masses, respectively

is performed at a fixed scale μ_r, known as the renormalization scale. As a result, α_S and the quark masses depend on the choice of μ_r. The short-distance mass $m(\mu_r)$ is referred to as $\overline{\text{MS}}$ mass, or running mass. However, since μ_r is an artificially-introduced scale, any physical observable calculated at all orders in perturbation theory does not depend on it. Considering a dimensionless observable $\mathcal{O}$ that depends on an energy scale Q, this fact can be expressed as:

$$\left[\mu^2\frac{\partial}{\partial\mu^2} + \beta(\alpha_S)\frac{\partial}{\partial\alpha_S} - \gamma_m(\alpha_S)\,m\frac{\partial}{\partial m}\right]\mathcal{O}\left(Q^2/\mu^2, \alpha_S, m/Q\right) = 0. \tag{2.13}$$

This expression defines the so-called renormalization group equations (RGEs):

$$\beta(\alpha_S) = \mu^2\frac{\partial\alpha_S}{\partial\mu^2}, \tag{2.14}$$

$$\mu^2\frac{\partial m}{\partial\mu^2} = -\gamma_m(\alpha_S)\,m, \tag{2.15}$$

which describe the scale dependence, or running, of α_S and the quark masses, respectively. The RGEs depend on the β function, $\beta(\alpha_S)$, and the mass anomalous dimension, $\gamma_m(\alpha_S)$. With this formalism, the dependence of the observable $\mathcal{O}$ on the physical scale Q is described in terms of the evolution of the QCD parameters with respect to the same scale.

If the value of α_S is small, Eqs. 2.14 and 2.15 can be solved in perturbation theory by expanding $\beta(\alpha_S)$ and $\gamma_m(\alpha_S)$ in powers of α_S:

$$\beta(\alpha_S) = -\alpha_S^2 \sum_{n=0}^{\infty} \beta_n \alpha_S^n, \tag{2.16}$$

$$\gamma_m(\alpha_S) = \alpha_S \sum_{n=0}^{\infty} c_n \alpha_S^n, \tag{2.17}$$

where the coefficients β_{n-1} and c_{n-1} contain the n-loop corrections to the propagators and vertices. An explicit solution of Eqs. 2.14 and 2.15 can be obtained at the lowest order in the expansion, i.e. at one-loop precision:

$$\alpha_S(\mu^2) = \frac{\alpha_S(\mu_0^2)}{1 + \beta_0 \alpha_S(\mu_0^2) \ln\left(\mu^2/\mu_0^2\right)}, \tag{2.18}$$

$$m(\mu) = m(\mu_0)\left[1 - c_0 \alpha_S(\mu) \ln\left(\frac{\mu^2}{\mu_0^2}\right)\right]. \tag{2.19}$$

Here, μ_0 represents the initial scale at which α_S and the quark masses are evaluated, $c_0 = 1/\pi$, and $\beta_0 = \left(33 - 2n_f\right)/12\pi$, where n_f is the number of active flavours, i.e. the number of quark flavours that can be considered massless at the energy scale at which the running is considered [10]. On the other hand, the values of $\alpha_S(\mu_0)$ and $m(\mu_0)$ cannot be calculated, and have to be determined experimentally. Currently, the β function and the anomalous dimension are known up to five-loop order [45–48].

The validity of the perturbative treatment of QCD at high energy is contained in Eq. 2.18, which implies that the value of the coupling constant decreases with increasing energy scales. For this reason, QCD is said to be an asymptotically-free theory [49, 50], which is a feature of all gauge theories based on non-Abelian symmetry groups in four space-time dimensions [51]. Equation 2.18 also implies that the value of α_S rapidly increases at low energy, which invalidates the perturbative treatment of QCD. The scale at which α_S diverges is referred to as Λ_{QCD}, and is of the order of 250 MeV [10]. This fact is at the origin of quark confinement, which is the property that quarks are always observed combined into colourless states called hadrons. Approximate solutions of the QCD Lagrangian at low energies can only be obtained by means of a non-perturbative approaches, such as lattice QCD [52].

Measuring the running of the QCD parameters is an important test of the validity of perturbative QCD, but also a tool for indirect searches for undiscovered physics. In fact, the RGEs can be modified by contributions from physics beyond the standard model, e.g. in the context of supersymmetric theories [53] or in scenarios where the fermion masses are dynamically generated [54]. The running of α_S has been verified at several experiments on a wide range of energy, as shown in Fig. 2.4. The running of the b quark mass was also demonstrated [55] using data from various experiments at the CERN LEP [56–59], SLAC SLC [60], and DESY HERA [61] colliders, and measurements of charm quark production in deep inelastic scattering at the DESY HERA were used to determine the running of the charm quark mass [62]. In all cases, good agreement with the RGE predictions is observed. The running of the top quark mass, instead, is investigated for the first time in this thesis.

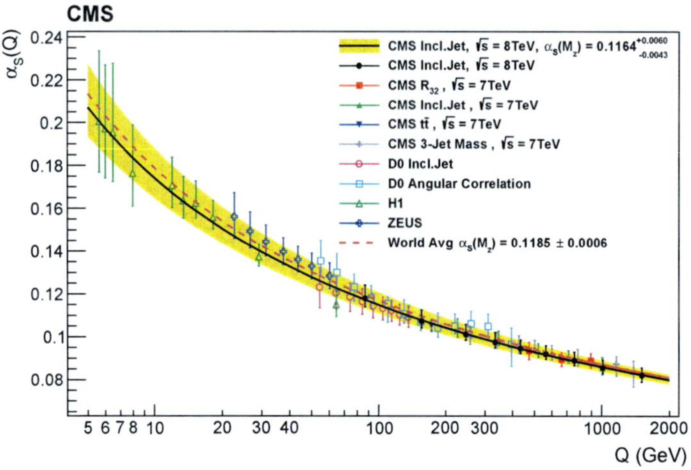

Fig. 2.4 Running of α_S determined using data from CMS and various other experiments at different colliders [63]

An alternative way to renormalise the quark masses is represented by the on-shell scheme, where the mass parameters are redefined order-by-order in perturbation theory so that they correspond to the pole in the quark propagator [10]. This approach is inspired by the renormalization procedure followed in quantum electrodynamics, and the corresponding mass is referred to as the pole mass. However, unlike leptons, quarks are subject to non-perturbative QCD effects that lead to a fundamental ambiguity in the quark pole mass definition of the order of Λ_{QCD}, known as renormalon ambiguity [64–66]. Nonetheless, the $\overline{\text{MS}}$ and pole masses can be related at every fixed order in perturbation theory. For example, at the one-loop level:

$$m_q(m_q) = m_q^{\text{pole}} \left[1 - \frac{4}{3\pi} \alpha_S(m_q) + \mathcal{O}(\alpha_S^2) \right], \tag{2.20}$$

where m_q^{pole} denotes the pole mass of quark q. The relation between the $\overline{\text{MS}}$ and the pole masses is currently known up to order α_S^4 [67].

Another common definition is the so-called MSR mass, which is a short-distance mass that depends on an infrared scale R [68]. The $\overline{\text{MS}}$ and pole masses can be obtained from the MSR mass for $R = m(\mu)$ and $R \to 0$, respectively. Like the $\overline{\text{MS}}$ mass, the MSR mass is not affected by renormalon ambiguity [68].

2.5 QCD Factorization and the Evolution of Parton Distribution Functions

In high-energy collisions involving hadrons, the large-momentum scattering is understood as the interaction between fundamental hadron constituents, called partons. In such collisions, the production cross section for any process can be calculated using a master formula known as QCD factorization theorem [69], which postulates the full factorization between short-distance (i.e. non-perturbative) and long-distance (i.e. perturbative) physics. The scale that defines the boundary between short- and long-distance dynamics is known as factorization scale, and is indicated with μ_f. According to the factorization theorem, the cross section for a proton-proton interaction of the kind pp $\to ab + X$ can be expressed as the convolution of the partonic cross section $\hat{\sigma}_{ij\to ab}$ with the parton distribution functions (PDFs) of the proton $f_i(x, \mu_\mathrm{f}^2)$:

$$\sigma_{\mathrm{pp}\to ab+X} = \sum_{i,j} \int_0^1 \mathrm{d}x_1 \mathrm{d}x_2 f_i(x_1, \mu_\mathrm{f}^2) f_j(x_2, \mu_\mathrm{f}^2) \hat{\sigma}_{ij\to ab} \left(x_1, x_2, s, \mu_\mathrm{f}^2, \mu_\mathrm{r}^2, \alpha_\mathrm{S}\right), \tag{2.21}$$

where $\sqrt{s}$ indicates the centre-of-mass energy of the collision. The PDF $f_i(x, \mu_\mathrm{f}^2)$ represent the probability that parton i, which can be either a quark or a gluon, carries a fraction x of the momentum of the proton. While the partonic cross section can be calculated in perturbation theory, the PDFs are universal non-perturbative quantities that must be determined experimentally.

The factorization scale is introduced beyond leading order in perturbation theory in order to remove infrared divergences arising from hard collinear initial-state radiation, and represents the minimum transverse momentum for which a collinear splitting of an interacting parton can be resolved. Therefore, the scale μ_f depends on the physical scale of the process. The dependence of the PDFs on the scale μ_f, also known as PDF evolution, is described by a set of differential equations known as the Dokshitzer-Gribov-Lipatov-Altarelli-Parisi (DGLAP) equations [70–72], which can be written in the form:

$$\mu^2 \frac{\partial}{\partial \mu^2} \begin{pmatrix} f_{q_i}(x, \mu^2) \\ f_g(x, \mu^2) \end{pmatrix} = \frac{\alpha_\mathrm{S}(\mu^2)}{2\pi} \sum_{q_j, \bar{q}_j} \int_x^1 \frac{\mathrm{d}y}{y} \times \begin{pmatrix} P_{q_i q_j}\left(\frac{x}{y}, \alpha_\mathrm{S}(\mu^2)\right) & P_{q_i g}\left(\frac{x}{y}, \alpha_\mathrm{S}(\mu^2)\right) \\ P_{g q_j}\left(\frac{x}{y}, \alpha_\mathrm{S}(\mu^2)\right) & P_{gg}\left(\frac{x}{y}, \alpha_\mathrm{S}(\mu^2)\right) \end{pmatrix} \begin{pmatrix} f_{q_j}(y, \mu^2) \\ f_g(y, \mu^2) \end{pmatrix}. \tag{2.22}$$

The functions $P_{ij}(z)$, known as the Altarelli-Parisi splitting functions, represent the probability for parton i to emit a parton j carrying a fraction z of its longitudinal momentum, in the collinear approximation. The splitting functions can be calculated in perturbation theory based on the Feynman diagrams shown in Fig. 2.5, and are currently known at up to three loops in QCD [73, 74]. Experimentally, PDFs are

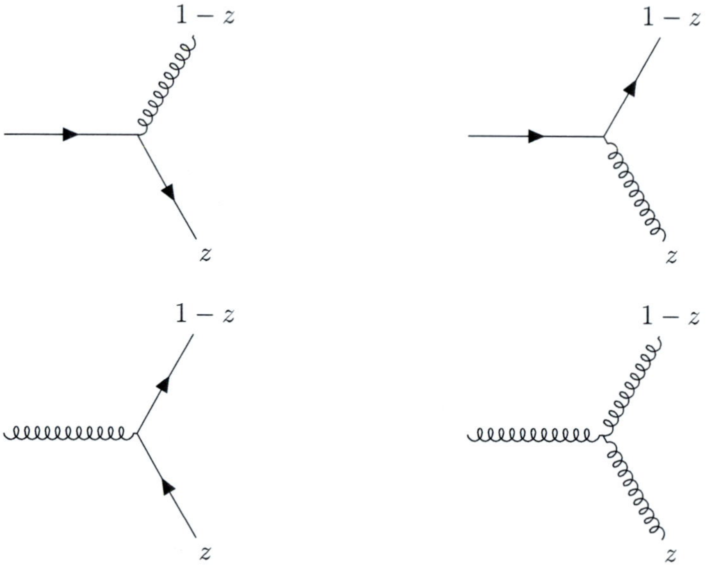

Fig. 2.5 Feynman diagrams for the processes corresponding to the Altarelli-Parisi splitting functions $P_{qq}(z)$ (upper left), $P_{gq}(z)$ (upper right), $P_{qg}(z)$ (lower left), and $P_{gg}(z)$ (lower right).

determined from data at a fixed reference scale, and are then evolved using DGLAP equations.

A precise determination of the PDFs has been performed using deep inelastic scattering (DIS) data from the DESY HERA [75] with further constraints from neutrino-nucleon scattering [76] and hadron collision data. Data collected by the experiments at the LHC can also provide additional constraints, especially on the gluon PDF [77–80]. Furthermore, the excellent agreement between the predicted and observed PDF evolution demonstrated using DIS data represents another important proof of the validity of perturbative QCD [75].

2.6 Monte Carlo Simulation

Experimental analyses often need realistic simulated events that can be directly compared to data. Such pseudo-events can be obtained using Monte Carlo (MC) event generators, which are tools that can be used to simulate the QCD dynamics at the different energy scales of an event. While partonic scattering is calculated in perturbation theory using numerical methods, the subsequent emission of soft and collinear partons is simulated using parton showers, which are programs designed to approximate higher-order terms with a perturbative approach. On the other hand, soft processes such as multiple parton interactions, the hadronization, and hadron decays can only

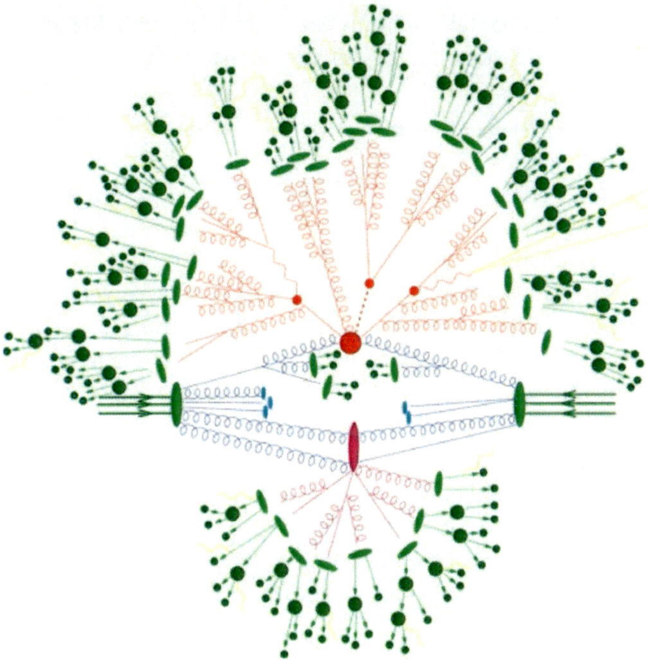

Fig. 2.6 Schematic representation of the MC simulation of a hadron-hadron collision. The red blob in the centre represents the large-momentum partonic interaction, surrounded by parton branchings simulated by the parton shower. The purple blob indicates a secondary scattering event, representing multiple-parton interactions. The hadronization is represented by light green blobs, while dark green blobs indicate the decays of final state hadrons [81]

be described with phenomenological models. A schematic representation of the MC simulation of a typical hadron-hadron collision event is shown in Fig. 2.6.

Parton showers rely on the same factorization properties of QCD that are used to describe the PDF evolution. Given a process with n final state partons and cross section σ_n, the cross section of the same process accompanied by the branching of parton i which generates an additional parton j can be approximated in the collinear limit as [82]:

$$d\sigma_{n+1} = \sigma_n \sum_{j,i} \frac{\alpha_S}{2\pi} \frac{\mathrm{d}k^2}{k^2} P^T_{ji}(z)\mathrm{d}z, \tag{2.23}$$

where z is the momentum fraction of the emitted parton, $P^T_{ij}(z)$ are the time-like splitting kernels related to the splitting functions introduced in Sect. 2.5, and k is a variable proportional to the relative transverse momentum of the emitted parton with respect to the parent parton. Equation 2.23 can be solved iteratively starting from the matrix element of the partonic scattering, adding a final-state branching at each iteration.

The soft and collinear divergences in Eq. 2.23, corresponding to $k \to 0$, can be removed by introducing a resolution parameter Q_0 which represents the scale below which a splitting is not resolved. By treating k as an ordering variable, the probability that there are no resolvable branchings with $k^2 > q^2$ is given by

$$\Delta_i(Q^2, q^2) = \exp\left[-\sum_j \int_{q^2}^{Q^2} \frac{dk^2}{k^2}\frac{\alpha_S}{2\pi}\int_{Q_0^2/k^2}^{1-Q_0^2/k^2} P_{ji}^T(z)\,dz\right], \tag{2.24}$$

where Q^2 is the shower starting scale. The function $\Delta_i(Q^2, q^2)$ is known as Sudakov form factor [83, 84]. In a parton shower, a value of q^2 is randomly generated according to the probability distribution of the first branching, given by the derivative of $\Delta_i(Q^2, q^2)$ with respect to q^2. If the generated value of q^2 is larger than Q_0^2, a splitting with $k^2 = q^2$ is generated according to $P_{ji}^T(z)$. Otherwise, the shower terminates. The same procedure can be applied to incoming partons to generate initial-state showers. In this case, Eq. 2.24 has to be modified in order to take into account the PDF evolution and the correct momentum fraction of the interacting parton [82].

The Sudakov form factor resums all powers of the largest logarithmic terms in the perturbative expansion, i.e. those of the kind $\left[\alpha_S \ln^2\left(Q^2/Q_0^2\right)\right]^m$. For this reason, the parton shower is said to be a leading-logarithmic approximation. Furthermore, the large logarithms that appear in the splitting kernels are resummed to all orders in the running of the coupling constant, $\alpha_S(q)$. However, as q^2 decreases during the shower evolution, $\alpha_S(q)$ becomes larger and the branching probabilities significantly increase. In order to ensure the validity of the perturbative approach, it is required that $Q_0^2 \gg \Lambda_{QCD}$. As a result, the choice of the cut-off parameter Q_0 can have a non-negligible impact on the resulting dynamics [82].

In order to avoid double counting of real emissions, a matching between the matrix element calculation and the parton shower is required. Since matrix element calculations can accurately describe large-angle, hard emissions, while parton showers are designed to simulate soft collinear radiation, the matching procedure consists in correcting the hardest emission from the parton shower. One of the most common matching procedure is the POWHEG method [85, 86], which consists in applying suitable corrections to the Sudakov form factor. The scale h_{damp} at which the matching is performed is known as resummation scale [87], and its value is tuned using data [88]. The matching procedure preserves both the NLO accuracy of the matrix element and the leading-logarithmic resummation properties of the parton shower.

The aspects of MC simulation discussed above are related to the large-momentum partonic scattering and the subsequent parton branchings. However, in hadron-hadron collisions, multiple parton-parton interaction occur at every event. These interactions give rise to soft QCD processes, known as underlying event, which cannot be calculated in perturbation theory and are simulated using phenomenological models that are directly tuned to the data [88]. In addition, the colour correlations between final state partons can significantly affect their kinematics [89]. Different colour reconnection models have been developed [90, 91], and are currently being tested using data from the LHC.

Finally, the partons generated at scales below the parton-shower cut-off scale are combined to form hadrons. Since hadronization occurs in the non-perturbative regime of QCD, this part of the process can only be simulated using phenomenological models. One of the most common is the so-called string model [92], which assumes that quark-antiquark pairs are subjects to a linearly rising potential compressed in a string. If the energy stored in the string is larger than that of a quark-antiquark pair, a $q\bar{q}$ pair is produced and the string fragments, until stable hadrons are formed. In this model, gluons are treated as kinks in the string, and can affect the angular distributions of the final state hadrons. A better phenomenological description of the fragmentation of heavy quarks is provided by the Bowler–Lund parametrization [93], which is an extension of the string model where the string potential is modified by the effect of the heavy quark mass. This results in a parametrization of the heavy-quark fragmentation function, i.e. the distribution of the fraction of the b quark momentum carried by the b hadron, in terms of two tunable parameters. An alternative parametrization is represented by the Peterson fragmentation function [94], which instead depends on a single parameter.

2.7 The Top Quark

The top quark was discovered in 1995 by the CDF and D0 Collaborations at the Fermilab Tevatron, using $p\bar{p}$ collisions at $\sqrt{s} = 1.8\,\text{TeV}$ [95, 96]. With a pole mass of $173.1 \pm 0.9\,\text{GeV}$ [10], the top quark is the most massive elementary particle currently known. For this reason, the decay time of the top quark is considerably smaller than the time scale of hadronization, and top quarks decay before forming bound states. This provides the possibility to study the properties of an unconfined quark, e.g. its polarization and charge [97, 98], and to directly measure the mass of the top quark from the invariant mass of its decay products. Furthermore, the value m_{t} provides a natural hard scale for testing the validity of perturbative QCD, and is also of particular importance in the context of electroweak physics. In fact, large corrections to the Higgs boson mass arise from diagrams such as the one shown in Fig. 2.7, which quadratically depend on the top Yukawa coupling, and therefore on m_{t} [99]. The first-order correction to the parameter μ^2 in the Higgs potential (Eq. 2.6) is calculated to be:

$$\delta\mu^2 = -\frac{3}{4\pi^2} m_{\mathrm{t}}^2 \frac{\Lambda^2}{v^2}, \tag{2.25}$$

where $\Lambda \gg 1\,\text{TeV}$ is the cut-off scale of the theory [99]. In order to obtain the correct physical Higgs mass, an unnatural fine-tuning of the tree-level parameter μ is required [99]. This fact is also known as the hierarchy (or naturalness) problem. Similarly, diagrams involving top-quark loops are responsible for the quantum corrections to the parameter λ in the Higgs potential. Therefore, the evolution of λ as a function of the energy scale depends on the values of α_{S}, m_{t}, and (to lesser extent) the Higgs boson mass [100–102], as shown in Fig. 2.8 (left). This result shows that,

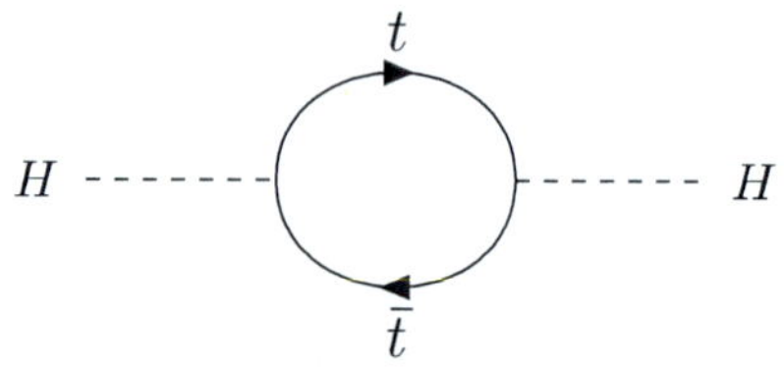

Fig. 2.7 Diagram representing the one-loop correction to the Higgs mass due to a top quark loop

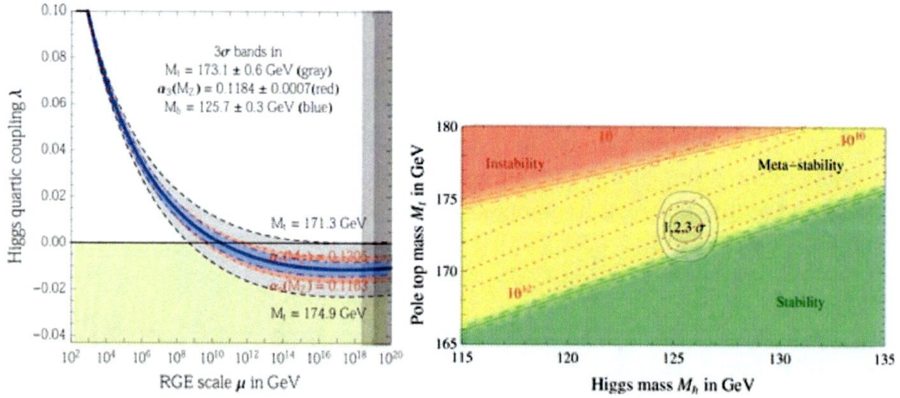

Fig. 2.8 Left: evolution of the Higgs quartic coupling as a function of the RGE scale, including the uncertainty in the values of α_S, m_t^{pole}, and the Higgs boson mass. Right: stability, meta-stability, and instability of the electroweak vacuum as a function of the top quark and Higgs boson masses, compared to the current knowledge of these two quantities [101]

at higher energy scales, the sign of λ, which is responsible for the stability of the electroweak vacuum, crucially depends on the value of m_t. With the current level of precision on m_t and the Higgs boson mass, only the unstable vacuum hypothesis can be confidently ruled out [100–102], as shown in Fig. 2.8 (right).

In pp collisions, top quarks are mainly produced as quark-antiquark ($t\bar{t}$) pairs, as shown in Fig. 2.9. This process has been studied by the experiments at the LHC in a wide range of energies, and is found to be well described in terms of perturbative QCD, as shown in Fig. 2.10. At $\sqrt{s} = 13$ TeV, the gluon fusion channel, which corresponds to the first two diagrams in Fig. 2.9, accounts for about 90% of the $t\bar{t}$ production [10]. With a total cross section of about 830 pb [10], $t\bar{t}$ production is one of the most abundant among the processes of interest for the LHC physics programme.

Top quark decays proceed via charged-current weak interaction, and the main decay channel is represented by the process $t \to Wb$. The branching ratio for this channel can be calculated in terms of the elements of the CKM matrix:

$$B_{bW} = \frac{|V_{tb}|^2}{|V_{tb}|^2 + |V_{ts}|^2 + |V_{td}|^2}, \tag{2.26}$$

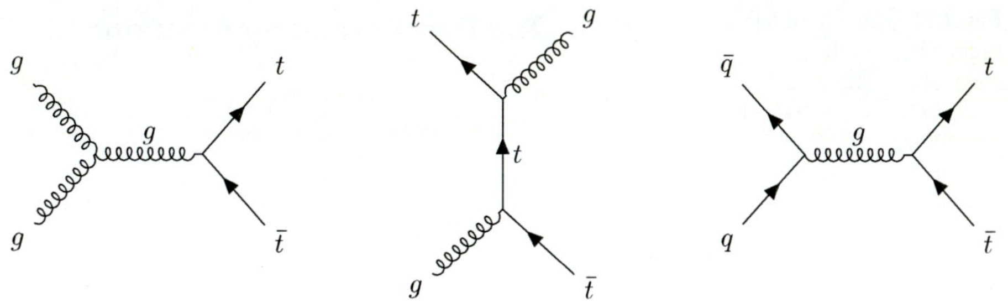

Fig. 2.9 Leading-order Feynman diagrams for $t\bar{t}$ production in hadronic collisions. The first two diagrams represent the gluon-fusion production mode in the s-channel and t-channel, respectively, while the diagram on the right-hand side represent the production mode corresponding to quark-antiquark annihilation

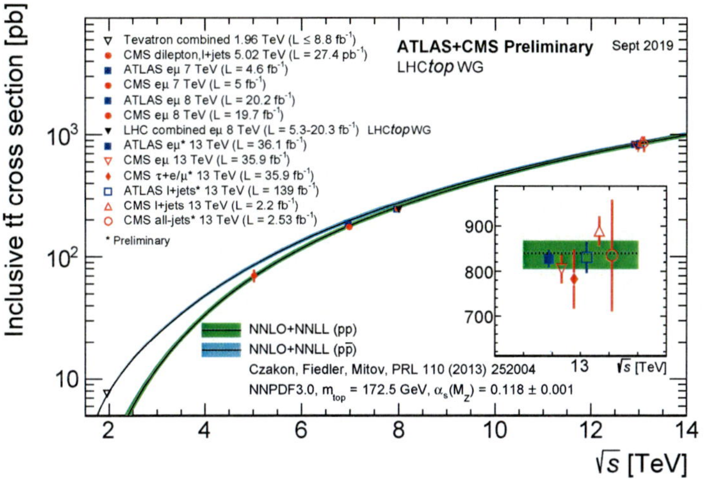

Fig. 2.10 Inclusive top quark pair production cross section as a function of the centre-of-mass energy in $p\bar{p}$ and pp collisions [103]. Measurement from the ATLAS and CMS Collaborations and from the experiments at the Fermilab Tevatron are compared to theoretical predictions in perturbative QCD obtained with the TOP++ program [104]

where V_{tq} is the CKM element that controls the $t \rightarrow Wq$ vertex. In a recent CMS measurement, this quantity was determined to be larger than 0.955 at 95% confidence level [105], while the SM prediction in terms of the measured CKM matrix elements is $B_{bW} = 0.998$ [10]. Based on the decays of the W bosons, $t\bar{t}$ final states can be classified into three different channels: the dileptonic, corresponding to events in which both W bosons decay into a lepton and a neutrino; the semileptonic, in which one W boson decays leptonically while the other one decays to a quark-antiquark pair; and the fully-hadronic, in which both W bosons decay hadronically. The branching ratios of the different channels are summarized in Fig. 2.11. Despite the low branching ratio, dileptonic channels are widely used in physics analyses since they can be experimentally identified with high purity. On the opposite, the fully hadronic channel

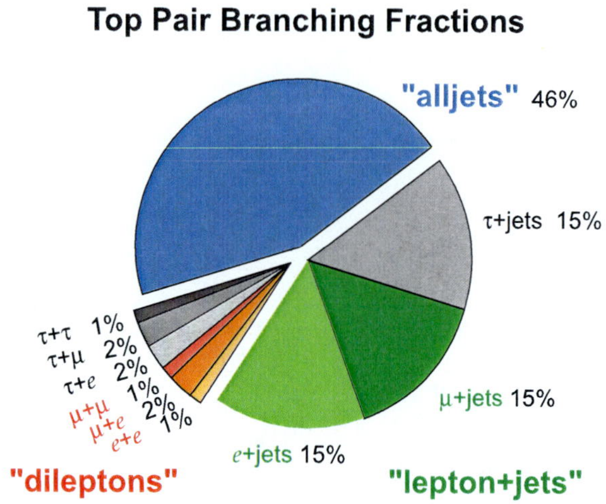

Fig. 2.11 Branching ratios of the different decay channels of a $t\bar{t}$ pair [106]. The semileptonic channels are indicated with "lepton+jets", while the fully-hadronic channel is referred to as "alljets"

accounts for almost half of the $t\bar{t}$ decays, but it is difficult to isolate from QCD multi-jet production. The semileptonic channels have intermediate properties, with moderate background contamination and significant branching ratios. The results presented in this thesis are obtained using $t\bar{t}$ events in the $e^{\mp}\mu^{\pm}$ final state, which corresponds to a branching ratio of about 2% [10].

Measurements of the top quark mass based on the reconstruction of the invariant mass of its decay products, often referred to as direct mass measurements, can reach a remarkable precision of a few per mille, as shown in Fig. 2.12. However, these measurements heavily rely on distributions derived from MC simulation, and are therefore sensitive to the details of the MC generator used [82, 107]. For this reason, the value of m_t determined with this method lacks a clear theoretical interpretation, and is understood as the value of the top quark mass parameter in the MC simulation (m_t^{MC}) that best describes the data. Dedicated studies suggest that the difference between m_t^{MC} and m_t^{pole} can be of the order of 1 GeV, and that the value of m_t^{MC} corresponds to that of $m_t^{MSR}(R)$, with R of the order of the cut-off scale in the parton shower [108]. The most recent measurement of m_t^{MC} in the dileptonic channel (Fig. 2.12) was developed for the scope of this thesis, and is presented in detail in Chap. 5.

Alternatively, the top quark mass can be determined in a well-defined renormalization scheme, e.g. on-shell or $\overline{MS}$, by comparing fixed-order theoretical predictions of observables sensitive to m_t to the corresponding measurements. Typical observables are the inclusive $t\bar{t}$ cross section, $\sigma_{t\bar{t}}$, or differential cross sections as a function of a sensitive variable, e.g. the invariant mass of the $t\bar{t}$ pair or that of the $t\bar{t}$ +1 jet system [109]. Similar methods can be used to determine the strong coupling constant and to provide additional constraints to the PDFs, e.g. as in Refs. [79, 110]. The most recent measurements of the top quark pole mass from the ATLAS and CMS Collaborations are shown in Fig. 2.13. The most recent NNLO result for m_t^{pole}

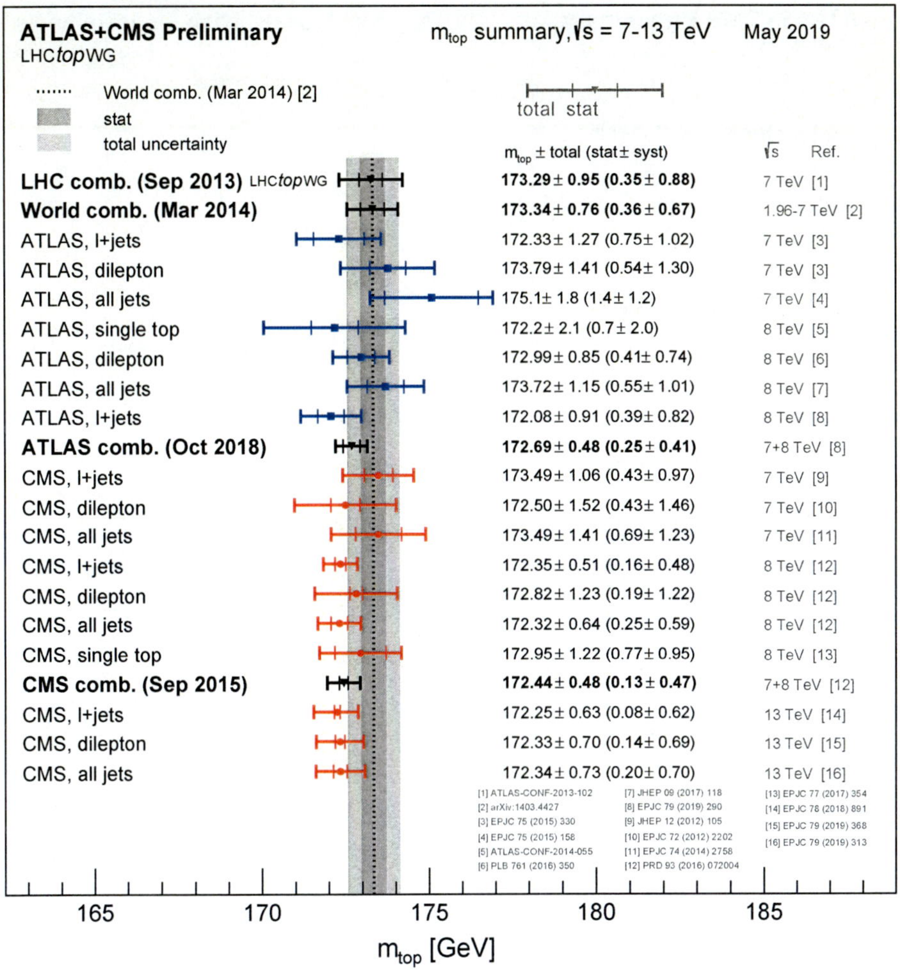

Fig. 2.12 Direct measurements of the top quark mass by the ATLAS and CMS Collaborations [103]

in Fig. 2.13 was obtained for the purpose of this thesis (Chap. 6), and is determined from a measurement of the inclusive $\mathrm{t\bar{t}}$ cross section at $\sqrt{s} = 13$ TeV (Chap. 5). The same measurement is used to extract the values of $\alpha_S(m_Z)$ and $m_t(m_t)$ at NNLO. A measurement of $m_t(m_t)$ at NLO has also been performed by the ATLAS Collaboration using $\mathrm{t\bar{t}}$+1 jet events at $\sqrt{s} = 8$ TeV [111], exploiting the observable proposed in Ref. [109].

The value of m_t can also be indirectly determined by means of so-called electroweak fits. These fits take as inputs experimental results, theoretical calculations, and model-dependent relations between the SM parameters. For example, the value of the top quark mass can affect the W boson propagator and the $Z \rightarrow \mathrm{b\bar{b}}$ vertex, as shown in Fig. 2.14, and can therefore be determined indirectly. Remarkably, the

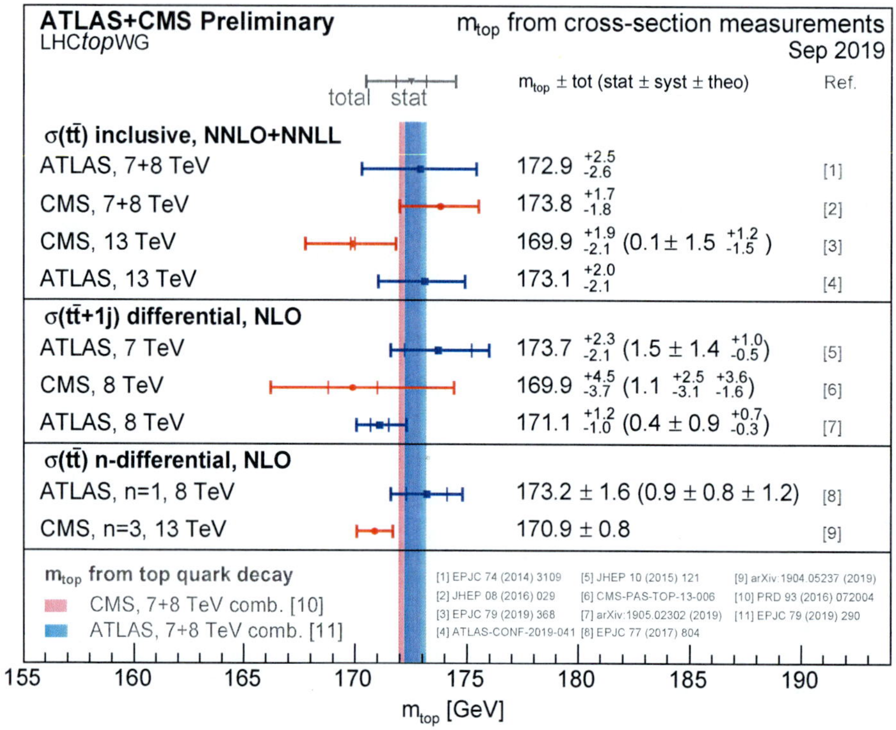

Fig. 2.13 Determinations of the top quark pole mass by the ATLAS and CMS Collaborations [103]

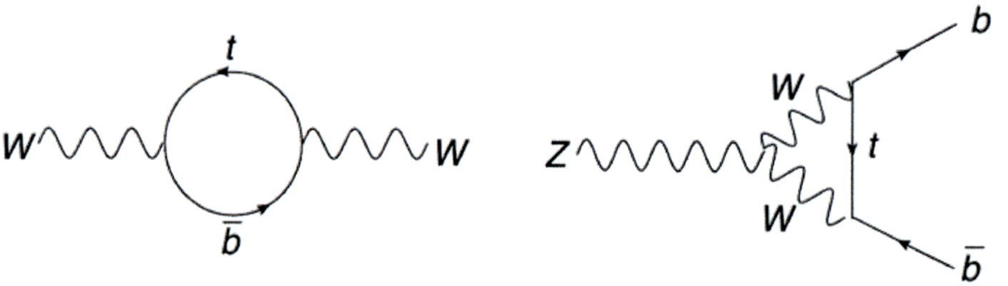

Fig. 2.14 Examples of one-loop corrections to the W boson propagator (left) and the $Z \to b\bar{b}$ vertex (right) involving the top quark [112]

value of m_t^{pole} obtained in recent electroweak fits is found to be in scarce agreement with those determined from $t\bar{t}$ data [112]. This fact further motivates the need of more precise measurements of m_t obtained with different methods in order to probe the limits of our current understanding of fundamental interactions.

2.8 Fixed Order Calculations for $\mathrm{t\bar{t}}$ Production

Precise calculations of the inclusive and differential $\mathrm{t\bar{t}}$ production cross sections can be performed at a fixed order in perturbation theory. As discussed in the previous section, these calculations allow m_{t} to be determined in a well-defined renormalization scheme, avoiding the ambiguities of the results of direct measurements. In particular, calculations in the $\overline{\mathrm{MS}}$ and on-shell schemes can be used to determine $m_{\mathrm{t}}(m_{\mathrm{t}})$ and $m_{\mathrm{t}}^{\mathrm{pole}}$, respectively.

In fixed-order calculations, large terms appear in the perturbative expansion of $\sigma_{\mathrm{t\bar{t}}}$ near the $\mathrm{t\bar{t}}$ production threshold [113]. For example, the emission of soft gluon gives rise to large logarithms of the kind $\alpha_{\mathrm{S}}^{k} \ln^{m} \beta$, where k is the loop order of the calculation, m is any integer number in the range $1 \leq m \leq 2k$, and $\beta = \sqrt{1 - 4m_{\mathrm{t}}^{2}/\hat{s}}$ is the parton velocity, a quantity that vanishes near the $\mathrm{t\bar{t}}$ production threshold. To date, these logarithms can be resummed in perturbation theory up to next-to-next-to-leading logarithmic (NNLL) precision [113, 114]. This procedure is known as soft-gluon resummation. In addition, so-called Coulomb terms arise from the exchange of gluons between non-relativistic top quark pairs. The corresponding terms are of the form $(\alpha_{\mathrm{S}}/\beta)^{k}$, and are automatically included in fixed-order calculations. However, it is also possible to resum the Coulomb terms simultaneously with the soft-gluon logarithms [115].

Theoretical predictions for the inclusive $\mathrm{t\bar{t}}$ cross section in the $\overline{\mathrm{MS}}$ scheme are obtained at next-to-next-to-leading order (NNLO) using the HATHOR 2.0 software [116], while NNLO calculations in the on-shell scheme are performed with the TOP++ 2.0 program [104], which also includes the soft gluon resummation at NNLL precision. The calculations are interfaced with different NNLO PDF sets, as discussed in detail in Chap. 6. Theoretical predictions for the differential $\mathrm{t\bar{t}}$ cross sections in the $\overline{\mathrm{MS}}$ scheme are currently only available at NLO precision. The calculations used in this thesis are implemented in a dedicated version [117] of the MCFM v6.8 program [118, 119]. All the above calculations are performed in the five-flavour number scheme, where all quarks except the top are considered to be massless. The same scheme is adopted in the considered PDFs, which means that the contribution from initial-state b quarks is absorbed into the PDFs.

A common procedure used to estimate the effect of missing higher order corrections consists in performing variations of the renormalization and factorization scale by a factor of two. In fact, such variations are known to generate terms of one order higher in α_{S} with respect to the order of the calculation. The validity of this approach is based on the comparison between the obtained uncertainties and calculations at different orders. In the context of $\mathrm{t\bar{t}}$ production, it was demonstrated that calculations of $\sigma_{\mathrm{t\bar{t}}}$ performed in the $\overline{\mathrm{MS}}$ scheme exhibit better perturbative convergence compared to the corresponding calculations in the on-shell scheme [117]. In fact, as shown in Fig. 2.15 for $\sqrt{s} = 8\,\mathrm{TeV}$, smaller differences between LO, NLO, and NNLO are observed in results obtained in the $\overline{\mathrm{MS}}$ scheme. Furthermore, the scale dependence of the calculated $\sigma_{\mathrm{t\bar{t}}}$ is found to be milder compared to the on-shell scheme. However, similar theoretical uncertainties are obtained with the two schemes when performing differential calculations [117].

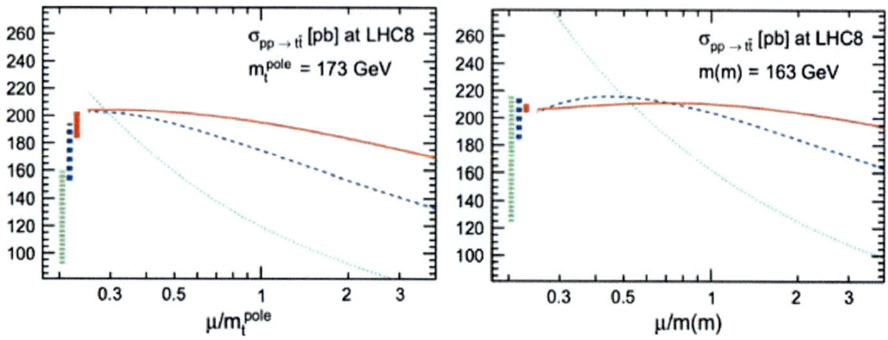

Fig. 2.15 Scale dependence of the LO (green), NLO (blue), and NNLO (red) predictions of the inclusive $\mathrm{t\bar{t}}$ cross section at $\sqrt{s} = 8\,\mathrm{TeV}$ in the on-shell (left) and $\overline{\mathrm{MS}}$ (right) schemes. The vertical bars represent the predicted values of $\sigma_{\mathrm{t\bar{t}}}$ and their uncertainty for $m_{\mathrm{t}}^{\mathrm{pole}} = 173\,\mathrm{GeV}$ and $m_{\mathrm{t}}(m_{\mathrm{t}}) = 163\,\mathrm{GeV}$, respectively [117]

2.9 Simulation of $\mathrm{t\bar{t}}$ Events

The measurements presented in this thesis make use of samples of simulated $\mathrm{t\bar{t}}$ events. The samples are generated using the POWHEG v2 [85, 86, 120, 121] NLO MC generator interfaced to the NNPDF3.0 NLO PDF set [122]. In the matrix-element calculation, the renormalization and factorization scales are chosen to be the transverse mass of the top quark, defined as $m_{\mathrm{T}} = \sqrt{m_{\mathrm{t}}^2 + p_{\mathrm{T}}^2}$. In the nominal simulation the value of the top quark mass is set to $m_{\mathrm{t}} = m_{\mathrm{t}}^{\mathrm{MC}} = 172.5\,\mathrm{GeV}$. A value of $\alpha_S(m_Z) = 0.118$, corresponding to the value utilized in the determination of the PDFs, is used in the calculation, where its evolution is calculated at two loops. The width of the top quark is set to the SM expectation of 1.31 GeV, and top quark decays are simulated at leading-order precision. The initial- and final-state showers are simulated using the PYTHIA 8.2 [123] program with the CUETP8M2T4 underlying event tune [88, 124], in which the best-fit values for h_{damp} and α_S are determined to be $h_{\mathrm{damp}} = 1.58\,^{+0.66}_{-0.59} \times m_{\mathrm{t}}$ and $\alpha_S(m_Z) = 0.111\,^{+0.015}_{-0.014}$, respectively [124]. The same program is also used to simulate the hadronization of quarks using a string model, while the fragmentation of heavy quarks is parametrized using the Bowler–Lund function. Since the parton shower provides a leading-logarithmic approximation of higher orders in QCD, the running of α_S and the PDF evolution in PYTHIA are calculated at one-loop precision. The starting scales for ISR and FSR showering are both set to m_{T}, and the cut-off scale is chosen to be 0.8 GeV. Similar setups are used for the simulation of the main background processes, which are tW [125] and Drell–Yan production [126].

References

1. Boyle L (2019) Standard model of particle physics, most complete diagram. https://commons.m.wikimedia.org/wiki/File:Standard_Model_Of_Particle_Physics--Most_Complete_Diagram.png. Accessed: November 2019
2. Wu CS, Ambler E, Hayward RW, Hoppes DD, Hudson RP (1957) Experimental test of parity conservation in beta decay. Phys Rev 105:1413. https://doi.org/10.1103/PhysRev.105.1413
3. Goldhaber M, Grodzins L, Sunyar AW (1958) Helicity of neutrinos. Phys Rev 109:1015. https://doi.org/10.1103/PhysRev.109.1015
4. ALEPH, DELPHI, L3, OPAL, and SLD Collaborations, LEP Electroweak Working Group, SLD Electroweak Group, SLD Heavy Flavour Group (2006) Precision electroweak measurements on the Z resonance. Phys Rept 427:257. https://doi.org/10.1016/j.physrep.2005.12.006, arXiv:hep-ex/0509008
5. Glashow SL (1961) Partial symmetries of weak interactions. Nucl Phys 22:579. https://doi.org/10.1016/0029-5582(61)90469-2
6. Weinberg S (1967) A model of leptons. Phys Rev Lett 19:1264. https://doi.org/10.1103/PhysRevLett.19.1264
7. Salam A (1968) Weak and electromagnetic interactions. Conf Proc C680519:367. https://doi.org/10.1142/9789812795915_0034
8. Higgs PW (1964) Broken symmetries and the masses of gauge bosons. Phys Rev Lett 13:508. https://doi.org/10.1103/PhysRevLett.13.508
9. Englert F, Brout R (1964) Broken symmetry and the mass of gauge vector mesons. Phys Rev Lett 13:321. https://doi.org/10.1103/PhysRevLett.13.321
10. Particle Data Group, Tanabashi M, et al (2018) Review of particle physics. Phys Rev D 98:030001. https://doi.org/10.1103/PhysRevD.98.030001
11. ATLAS Collaboration (2012) Observation of a new particle in the search for the standard model Higgs boson with the ATLAS detector at the LHC. Phys Lett B 716:1. https://doi.org/10.1016/j.physletb.2012.08.020, arXiv:1207.7214
12. Collaboration CMS (2012) Observation of a new boson at a mass of 125 GeV with the CMS experiment at the LHC. Phys Lett B 716:30. https://doi.org/10.1016/j.physletb.2012.08.021 arXiv:1207.7235
13. ATLAS, CMS Collaboration (2016) Measurements of the Higgs boson production and decay rates and constraints on its couplings from a combined ATLAS and CMS analysis of the LHC pp collision data at $\sqrt{s} = 7$ and 8 TeV. JHEP 08:045. https://doi.org/10.1007/JHEP08(2016)045, arXiv:1606.02266
14. ATLAS Collaboration Combined measurements of Higgs boson production and decay using up to 80 fb^{-1} of proton–proton collision data at $\sqrt{s} = 13$ TeV collected with the ATLAS experiment
15. ATLAS Collaboration (2018) Observation of Higgs boson production in association with a top quark pair at the LHC with the ATLAS detector. Phys Lett B 784:173. https://doi.org/10.1016/j.physletb.2018.07.035, arXiv:1806.00425
16. CMS Collaboration (2018) Observation of $\mathrm{t\bar{t}H}$ production. Phys Rev Lett 120. https://doi.org/10.1103/PhysRevLett.120.231801, arXiv:1804.02610
17. ATLAS Collaboration (2018) Observation of $\mathrm{H} \rightarrow \mathrm{b\bar{b}}$ decays and VH production with the ATLAS detector. Phys Lett B 786:59. https://doi.org/10.1016/j.physletb.2018.09.013, arXiv:1808.08238
18. CMS Collaboration (2018) Observation of Higgs boson decay to bottom quarks. Phys Rev Lett 121. https://doi.org/10.1103/PhysRevLett.121.121801, arXiv:1808.08242
19. ATLAS Collaboration (2019) Cross-section measurements of the Higgs boson decaying into a pair of τ-leptons in proton-proton collisions at $\sqrt{s} = 13$ TeV with the ATLAS detector. Phys Rev D 99:072001. https://doi.org/10.1103/PhysRevD.99.072001, arXiv:1811.08856
20. CMS Collaboration (2018) Observation of the Higgs boson decay to a pair of τ leptons with the CMS detector. Phys Lett B 779:283. https://doi.org/10.1016/j.physletb.2018.02.004, arXiv:1708.00373

21. ATLAS Collaboration (2019) A search for the dimuon decay of the standard model Higgs boson in pp collisions at $\sqrt{s} = 13$ TeV with the ATLAS detector. In: ATLAS conference note ATLAS-CONF-2019-028, CERN, Geneva. http://cds.cern.ch/record/2682155
22. CMS Collaboration (2019) Search for the Higgs boson decaying to two muons in proton-proton collisions at $\sqrt{s} = 13$ TeV. Phys Rev Lett 122. https://doi.org/10.1103/PhysRevLett.122.021801, arXiv:1807.06325
23. ATLAS Collaboration (2018) Search for the decay of the Higgs boson to charm quarks with the ATLAS experiment. Phys Rev Lett 120:211802. https://doi.org/10.1103/PhysRevLett.120.211802, arXiv:1802.04329
24. CMS Collaboration (2019) Search for the standard model Higgs boson decaying to charm quarks, CMS physics analysis summary CMS-PAS-HIG-18-031, CERN, Geneva. https://cds.cern.ch/record/2682638
25. Cabibbo N (1963) Unitary symmetry and leptonic decays. Phys Rev Lett 10:531. https://doi.org/10.1103/PhysRevLett.10.531
26. Kobayashi M, Maskawa T (1973) CP violation in the renormalizable theory of weak interaction. Prog Theor Phys 49:652. https://doi.org/10.1143/PTP.49.652
27. UTfit Collaboration (2017) Latest results for the unitary triangle fit from the UTfit Collaboration. PoS CKM2016:096. https://doi.org/10.22323/1.291.0096
28. Collaboration Super-Kamiokande (1998) Evidence for oscillation of atmospheric neutrinos. Phys Rev Lett 81:1562. https://doi.org/10.1103/PhysRevLett.81.1562, arXiv:hep-ex/9807003
29. SNO Collaboration (2002) Direct evidence for neutrino flavor transformation from neutral current interactions in the Sudbury neutrino observatory. Phys Rev Lett 89. https://doi.org/10.1103/PhysRevLett.89.011301, arXiv:nucl-ex/0204008
30. OPERA Collaboration (2018) Final results of the OPERA experiment on ν_τ appearance in the CNGS neutrino beam. Phys Rev Lett 120:211801 https://doi.org/10.1103/PhysRevLett.121.139901,10.1103/PhysRevLett.120.211801, arXiv:1804.04912
31. Pontecorvo B (1958) Inverse beta processes and nonconservation of lepton charge. Sov Phys JETP 7:172
32. Maki Z, Nakagawa M, Sakata S (1962) Remarks on the unified model of elementary particles. Prog Theor Phys 28:870. https://doi.org/10.1143/PTP.28.870
33. Bilenky SM, Hošek J, Petcov ST (1980) On oscillations of neutrinos with Dirac and Majorana masses. Phys Lett B 94:495. https://doi.org/10.1016/0370-2693(80)90927-2
34. Esteban I, Gonzalez-Garcia MC, Maltoni M, Martinez-Soler I, Schwetz T (2017) Updated fit to three neutrino mixing: exploring the accelerator-reactor complementarity. JHEP 01:087. https://doi.org/10.1007/JHEP01(2017)087, arXiv:1611.01514
35. Dolinski MJ, Poon AWP, Rodejohann W Neutrinoless double-beta decay: status and prospects. arXiv:1902.04097
36. KATRIN Collaboration (2019) An improved upper limit on the neutrino mass from a direct kinematic method by KATRIN. Phys Rev Lett 123:221802. https://doi.org/10.1103/PhysRevLett.123.221802, arXiv:1909.06048
37. GERDA Collaboration (2018) Improved limit on neutrinoless double-β decay of ^{76}Ge from GERDA Phase II. Phys Rev Lett 120:132503. https://doi.org/10.1103/PhysRevLett.120.132503, arXiv:1803.11100
38. CUORE Collaboration (2018) First results from CUORE: a search for lepton number violation via $0\nu\beta\beta$ decay of ^{130}Te. Phys Rev Lett 120:132501. https://doi.org/10.1103/PhysRevLett.120.132501, arXiv:1710.07988
39. Collaboration KamLAND-Zen (2016) Search for Majorana neutrinos near the inverted mass hierarchy region with KamLAND-Zen. Phys Rev Lett 117. https://doi.org/10.1103/PhysRevLett.117.109903,10.1103/PhysRevLett.117.082503, arXiv:1605.02889
40. Gell-Mann M (1964) A schematic model of baryons and mesons. Phys Lett 8:214. https://doi.org/10.1016/S0031-9163(64)92001-3
41. Zweig G (1964) An SU_3 model for strong interaction symmetry and its breaking; Version 1, Technical report CERN-TH-401, CERN, Geneva. http://cds.cern.ch/record/352337

42. Zweig G (1964) An SU_3 model for strong interaction symmetry and its breaking; Version 2, Technical Report CERN-TH-412, CERN, Geneva. http://cds.cern.ch/record/570209
43. TASSO Collaboration (1979) Evidence for planar events in e^+e^- annihilation at high-energies. Phys Lett B 86:243. https://doi.org/10.1016/0370-2693(79)90830-X
44. 't Hooft G, Veltman MJG (1972) Regularization and renormalization of gauge fields. Nucl Phys B 44:189. https://doi.org/10.1016/0550-3213(72)90279-9
45. Baikov PA, Chetyrkin KG, Kühn JH (2017) Five-loop running of the QCD coupling constant. Phys Rev Lett 118. https://doi.org/10.1103/PhysRevLett.118.082002, arXiv:1606.08659
46. Luthe T, Maier A, Marquard P, Schröder Y (2017) Complete renormalization of QCD at five loops. JHEP 03:020. https://doi.org/10.1007/JHEP03(2017)020, arXiv:1701.07068
47. Baikov PA, Chetyrkin KG, Kühn JH (2014) Quark Mass and Field Anomalous Dimensions to $\mathcal{O}(\alpha_s^5)$. JHEP 10:076. https://doi.org/10.1007/JHEP10(2014)076, arXiv:1402.6611
48. Luthe T, Maier A, Marquard P, Schröder Y (2017) Five-loop quark mass and field anomalous dimensions for a general gauge group. JHEP 01:081. https://doi.org/10.1007/JHEP01(2017)081, arXiv:1612.05512
49. Gross DJ, Wilczek F (1973) Ultraviolet behavior of nonabelian gauge theories. Phys Rev Lett 30:1343. https://doi.org/10.1103/PhysRevLett.30.1343
50. Politzer HD (1973) Reliable perturbative results for strong interactions? Phys Rev Lett 30:1346. https://doi.org/10.1103/PhysRevLett.30.1346
51. Gross DJ, Wilczek F (1973) Asymptotically free gauge theories—I. Phys Rev D 8:3633. https://doi.org/10.1103/PhysRevD.8.3633
52. Wilson KG (1974) Confinement of quarks. Phys Rev D 10:2445. https://doi.org/10.1103/PhysRevD.10.2445
53. Mihaila L (2013) Precision calculations in supersymmetric theories. Adv High Energy Phys. https://doi.org/10.1155/2013/607807, arXiv:1310.6178
54. Christensen ND, Shrock R (2005) Implications of dynamical generation of standard-model fermion masses. Phys Rev Lett 94. https://doi.org/10.1103/PhysRevLett.94.241801, arXiv:hep-ph/0501294
55. Behnke O, Geiser A, Lisovyi M (2015) Charm, beauty and top at HERA. Prog Part Nucl Phys 84:1. https://doi.org/10.1016/j.ppnp.2015.06.002, arXiv:1506.07519
56. DELPHI Collaboration (2008) Study of b-quark mass effects in multijet topologies with the DELPHI detector at LEP. Eur Phys J C 55:525. https://doi.org/10.1140/epjc/s10052-008-0631-5, arXiv:0804.3883
57. DELPHI Collaboration (2006) Determination of the b quark mass at the M_Z scale with the DELPHI detector at LEP. Eur Phys J C 46:569. https://doi.org/10.1140/epjc/s2006-02497-6
58. ALEPH Collaboration (2000) A Measurement of the b quark mass from hadronic Z decays. Eur Phys J C 18:1. https://doi.org/10.1007/s100520000533, arXiv:hep-ex/0008013
59. OPAL Collaboration (2001) Determination of the b quark mass at the Z mass scale. Eur Phys J C 21:411. https://doi.org/10.1007/s100520100746, arXiv:hep-ex/0105046
60. Brandenburg A, Burrows PN, Muller D, Oishi N, Uwer P (1999) Measurement of the running b quark mass using $e^+e^- \to b\bar{b}g$ events. Phys Lett B 468:168. https://doi.org/10.1016/S0370-2693(99)01194-6, arXiv:hep-ph/9905495
61. ZEUS Collaboration (2014) Measurement of beauty and charm production in deep inelastic scattering at HERA and measurement of the beauty-quark mass. JHEP 09:127. https://doi.org/10.1007/JHEP09(2014)127, arXiv:1405.6915
62. Gizhko A, et al (2017) Running of the charm-quark mass from HERA deep-inelastic scattering data. Phys Lett B 775:233. https://doi.org/10.1016/j.physletb.2017.11.002, arXiv:1705.08863
63. CMS Collaboration (2019) Summary of α_S running. https://twiki.cern.ch/twiki/bin/view/CMSPublic/PhysicsResultsCombined#Summary_of_alphaS_running. Accessed: November 2019
64. Beneke M, Braun VM (1994) Heavy quark effective theory beyond perturbation theory: renormalons, the pole mass and the residual mass term. Nucl Phys B 426:301. https://doi.org/10.1016/0550-3213(94)90314-X, arXiv:hep-ph/9402364

65. Bigi IIY, Shifman MA, Uraltsev NG, Vainshtein AI (1994) The Pole mass of the heavy quark: perturbation theory and beyond. Phys Rev D 50:2234. https://doi.org/10.1103/PhysRevD.50.2234, arXiv:hep-ph/9402360
66. Smith MC, Willenbrock SS (1997) Top quark pole mass. Phys Rev Lett 79:3825. https://doi.org/10.1103/PhysRevLett.79.3825, arXiv:hep-ph/9612329
67. Marquard P, Smirnov AV, Smirnov VA, Steinhauser M (2015) Quark mass relations to four-loop order in perturbative QCD. Phys Rev Lett 114:142002. https://doi.org/10.1103/PhysRevLett.114.142002, arXiv:1502.01030
68. Hoang AH, Jain A, Scimemi I, Stewart IW (2008) Infrared renormalization group flow for heavy quark masses. Phys Rev Lett 101:151602. https://doi.org/10.1103/PhysRevLett.101.151602, arXiv:0803.4214
69. Collins JC, Soper DE, Sterman GF (1989) Factorization of hard processes in QCD. Adv Ser Direct High Energy Phys 5:1. https://doi.org/10.1142/9789814503266_0001, arXiv:hep-ph/0409313
70. Altarelli G, Parisi G (1977) Asymptotic freedom in parton language. Nucl Phys B 126:298. https://doi.org/10.1016/0550-3213(77)90384-4
71. Dokshitzer YL (1977) Calculation of the structure functions for deep inelastic scattering and e^+e^- annihilation by perturbation theory in quantum chromodynamics. Sov Phys JETP 46:641
72. Gribov VN, Lipatov LN (1972) Deep inelastic ep scattering in perturbation theory. Sov J Nucl Phys 15:438
73. Vogt A, Moch S, Vermaseren JAM (2004) The three-loop splitting functions in QCD: the singlet case. Nucl Phys B 691:129. https://doi.org/10.1016/j.nuclphysb.2004.04.024, arXiv:hep-ph/0404111
74. Moch S, Vermaseren JAM, Vogt A (2004) The three loop splitting functions in QCD: the nonsinglet case. Nucl Phys B 688:101. https://doi.org/10.1016/j.nuclphysb.2004.03.030, arXiv:hep-ph/0403192
75. H1 and ZEUS Collaborations (2015) Combination of measurements of inclusive deep inelastic $e^{\pm}p$ scattering cross sections and QCD analysis of HERA data. Eur Phys J C 75:580. https://doi.org/10.1140/epjc/s10052-015-3710-4, arXiv:1506.06042
76. NuTEV Collaboration (2006) Precise measurement of neutrino and anti-neutrino differential cross sections. Phys Rev D 74:012008. https://doi.org/10.1103/PhysRevD.74.012008, arXiv:hep-ex/0509010
77. NuTEV Collaboration (2006) Precise measurement of neutrino and anti-neutrino differential cross sections. Phys Rev D 74:012008. https://doi.org/10.1103/PhysRevD.74.012008, arXiv:hep-ex/0509010
78. Dulat S, Hou T, Gao J, Guzzi M, Huston J, Nadolsky P et al (2016) New parton distribution functions from a global analysis of quantum chromodynamics. Phys Rev D 93. https://doi.org/10.1103/PhysRevD.93.033006, arXiv:1506.07443
79. Alekhin S, Blümlein J, Moch S, Placakyte R (2017) Parton distribution functions, α_S, and heavy-quark masses for LHC Run II. Phys Rev D 96. https://doi.org/10.1103/PhysRevD.96.014011, arXiv:1701.05838
80. NNPDF Collaboration (2017) Parton distributions from high-precision collider data. Eur Phys J C 77:663. https://doi.org/10.1140/epjc/s10052-017-5199-5, arXiv:1706.00428
81. Höche S (2015) Introduction to parton-shower event generators. In: Proceedings, theoretical advanced study institute in elementary particle physics: journeys through the precision frontier: amplitudes for colliders (TASI 2014): Boulder, Colorado, p 235 (2015). arXiv:1411.4085
82. Buckley A, et al (2011) General-purpose event generators for LHC physics. Phys Rept 504:145. https://doi.org/10.1016/j.physrep.2011.03.005, arXiv:1101.2599
83. Sudakov VV (1956) Vertex parts at very high-energies in quantum electrodynamics. Sov Phys JETP 3:65
84. Collins JC (1989) Sudakov form-factors. Adv Ser Direct High Energy Phys 5:573. https://doi.org/10.1142/9789814503266_0006, arXiv:hep-ph/0312336
85. Nason P (2004) A New method for combining NLO QCD with shower Monte Carlo algorithms. JHEP 11:040. https://doi.org/10.1088/1126-6708/2004/11/040, arXiv:hep-ph/0409146

86. Frixione S, Nason P, Oleari C (2007) Matching NLO QCD computations with parton shower simulations: the POWHEG method. JHEP 11:070. https://doi.org/10.1088/1126-6708/2007/11/070, arXiv:0709.2092
87. Ježo T, Lindert JM, Moretti N, Pozzorini S (2018) New NLOPS predictions for $t\bar{t}$ + b-jet production at the LHC. Eur Phys J C 78:502. https://doi.org/10.1140/epjc/s10052-018-5956-0, arXiv:1802.00426
88. Skands P, Carrazza , Rojo J (2014) Tuning PYTHIA 8.1: the Monash 2013 tune. Eur Phys J C 74:3024. https://doi.org/10.1140/epjc/s10052-014-3024-y, arXiv:1404.5630
89. Gustafson G, Pettersson U, Zerwas PM (1988) Jet final states in WW pair production and color screening in the QCD vacuum. Phys Lett B 209:90. https://doi.org/10.1016/0370-2693(88)91836-9
90. Argyropoulos S, Sjöstrand T (2014) Effects of color reconnection on $t\bar{t}$ final states at the LHC. JHEP 11:043. https://doi.org/10.1007/JHEP11(2014)043, arXiv:1407.6653
91. Christiansen JR, Skands PZ (2015) String formation beyond leading colour. JHEP 08:003. https://doi.org/10.1007/JHEP08(2015)003, arXiv:1505.01681
92. Andersson B, Gustafson G, Ingelman G, Sjöstrand T (1983) Parton fragmentation and string dynamics. Phys Rept 97:31. https://doi.org/10.1016/0370-1573(83)90080-7
93. Bowler MG (1981) e^+e^- production of heavy quarks in the string model. Z Phys C 11:169. https://doi.org/10.1007/BF01574001
94. Peterson C, Schlatter D, Schmitt I, Zerwas PM (1983) Scaling violations in inclusive e^+e^- annihilation spectra. Phys Rev D 27:105. https://doi.org/10.1103/PhysRevD.27.105
95. Collaboration CDF (1995) Observation of top quark production in $\bar{p}p$ collisions. Phys Rev Lett 74:2626. https://doi.org/10.1103/PhysRevLett.74.2626, arXiv:hep-ex/9503002
96. D0 Collaboration (1995) Observation of the top quark. Phys Rev Lett 74:2632. https://doi.org/10.1103/PhysRevLett.74.2632, arXiv:hep-ex/9503003
97. Collaboration CMS (2019) Measurement of the top quark polarization and $t\bar{t}$ spin correlations using dilepton final states in proton-proton collisions at $\sqrt{s}$ = 13TeV. Phys Rev D 100:072002. https://doi.org/10.1103/PhysRevD.100.072002, arXiv:1907.03729
98. ATLAS Collaboration (2013) Measurement of the top quark charge in pp collisions at $\sqrt{s}$ = 7TeV with the ATLAS detector. JHEP 11:031. https://doi.org/10.1007/JHEP11(2013)031, arXiv:1307.4568
99. de Gouvêa A, Hernandez D, Tait TMP (2014) Criteria for natural hierarchies. Phys Rev D 89:115005. https://doi.org/10.1103/PhysRevD.89.115005, arXiv:1402.2658
100. Alekhin S, Djouadi A, Moch S (2012) The top quark and Higgs boson masses and the stability of the electroweak vacuum. Phys Lett B 716:214. https://doi.org/10.1016/j.physletb.2012.08.024, arXiv:1207.0980
101. Degrassi G, Di Vita S, Elias-Miró J, Espinosa JR, Giudice GF, Isidori G, et al (2012) Higgs mass and vacuum stability in the standard model at NNLO. JHEP 08:098. https://doi.org/10.1007/JHEP08(2012)098, arXiv:1205.6497
102. Bednyakov A, Kniehl B, Pikelner A, Veretin O (2015) Stability of the electroweak vacuum: gauge independence and advanced precision. Phys Rev Lett 115:201802. https://doi.org/10.1103/PhysRevLett.115.201802, arXiv:1507.08833
103. LHC Top Working Group (2019) LHCTopWG Summary Plots. https://twiki.cern.ch/twiki/bin/view/LHCPhysics/LHCTopWGSummaryPlots. Accessed: November 2019
104. Czakon M, Mitov A (2014) Top++: a program for the calculation of the top-pair cross-section at hadron colliders. Comput. Phys Commun 185:2930. https://doi.org/10.1016/j.cpc.2014.06.021, arXiv:1112.5675
105. Collaboration CMS (2014) Measurement of the ratio $\mathcal{B}(t \rightarrow Wb)/\mathcal{B}(t \rightarrow Wq)$ in pp collisions at $\sqrt{s}$ = 8 TeV. Phys Lett B 736:33. https://doi.org/10.1016/j.physletb.2014.06.076, arXiv:1404.2292
106. D0 Collaboration (2019) Useful diagrams of top signals and backgrounds. https://www-d0.fnal.gov/Run2Physics/top/top_public_web_pages/top_feynman_diagrams.html. Accessed: December 2019

107. Hoang AH, Plätzer S, Samitz D (2018) On the cutoff dependence of the quark mass parameter in angular ordered parton showers. JHEP 10:200. https://doi.org/10.1007/JHEP10(2018)200, arXiv:1807.06617
108. Butenschön M, Dehnadi B, Hoang AH, Mateu V, Preisser M, Stewart IW (2016) Top quark mass calibration for Monte Carlo event generators. Phys Rev Lett 117:232001. https://doi.org/10.1103/PhysRevLett.117.232001, arXiv:1608.01318
109. Alioli S, Fernandez P, Fuster J, Irles A, Moch S, Uwer P, et al (2013) A new observable to measure the top-quark mass at hadron colliders. Eur Phys J C 73:2438. https://doi.org/10.1140/epjc/s10052-013-2438-2, arXiv:1303.6415
110. CMS Collaboration (2019) Measurement of $t\bar{t}$ normalised multi-differential cross sections in pp collisions at $\sqrt{s} = 13$ TeV, and simultaneous determination of the strong coupling strength, top quark pole mass, and parton distribution functions. Submitted to: Eur Phys J. arXiv:1904.05237
111. ATLAS Collaboration Measurement of the top-quark mass in $t\bar{t} + 1$-jet events collected with the ATLAS detector in pp collisions at $\sqrt{s} = 8$TeV. arXiv:1905.02302
112. Erler J, Schott M (2019) Electroweak precision tests of the standard model after the discovery of the Higgs boson. Prog Part Nucl Phys 106:68. https://doi.org/10.1016/j.ppnp.2019.02.007, arXiv:1902.05142
113. Beneke M, Falgari P, Schwinn C (2010) Soft radiation in heavy-particle pair production: all-order colour structure and two-loop anomalous dimension. Nucl Phys B 828:69. https://doi.org/10.1016/j.nuclphysb.2009.11.004, arXiv:0907.1443
114. Czakon M, Mitov A, Sterman GF (2009) Threshold resummation for top-pair hadroproduction to next-to-next-to-leading log. Phys Rev D 80:074017. https://doi.org/10.1103/PhysRevD.80.074017, arXiv:0907.1790
115. Piclum J, Schwinn C (2018) Soft-gluon and Coulomb corrections to hadronic top-quark pair production beyond NNLO. JHEP 03:164. https://doi.org/10.1007/JHEP03(2018)164, arXiv:1801.05788
116. Aliev M, Lacker H, Langenfeld U, Moch S, Uwer P, Wiedermann M (2011) Hathor: hadronic top and heavy quarks cross section calculatoR. Comput Phys Commun 182:1034. https://doi.org/10.1016/j.cpc.2010.12.040, arXiv:1007.1327
117. Dowling M, Moch S (2014) Differential distributions for top-quark hadro-production with a running mass. Eur Phys J C 74:3167. https://doi.org/10.1140/epjc/s10052-014-3167-x, arXiv:1305.6422
118. Campbell JM, Ellis RK (2010) MCFM for the Tevatron and the LHC. Nucl Phys B Proc Suppl 10:205–206. https://doi.org/10.1016/j.nuclphysbps.2010.08.011, arXiv:1007.3492
119. Campbell JM, Ellis RK (2015) Top-quark processes at NLO in production and decay. J Phys G 42:015005. https://doi.org/10.1088/0954-3899/42/1/015005, arXiv:1204.1513
120. Alioli S, Nason P, Oleari C, Re E (2010) A general framework for implementing NLO calculations in shower Monte Carlo programs: the Powheg BOX. JHEP 06:043. https://doi.org/10.1007/JHEP06(2010)043, arXiv:1002.2581
121. Frixione S, Nason P, Ridolfi G (2007) A positive-weight next-to-leading-order Monte Carlo for heavy flavour hadroproduction. JHEP 09:126. https://doi.org/10.1088/1126-6708/2007/09/126, arXiv:0707.3088
122. NNPDF Collaboration (2015) Parton distributions for the LHC Run II. JHEP 04:040. https://doi.org/10.1007/JHEP04(2015)040, arXiv:1410.8849
123. Sjöstrand T, Ask S, Christiansen JR, Corke R, Desai N, Ilten P, et al (2015) An introduction to PYTHIA 8.2. Comput Phys Commun 191:159. https://doi.org/10.1016/j.cpc.2015.01.024, arXiv:1410.3012
124. CMS Collaboration (2016) Investigations of the impact of the parton shower tuning in PYTHIA 8 in the modelling of $t\bar{t}$ at $\sqrt{s} = 8$ and 13TeV, CMS physics analysis summary CMS-PAS-TOP-16-021, CERN, Geneva. https://cds.cern.ch/record/2235192

125. Re E (2011) Single-top Wt-channel production matched with parton showers using the Powheg method. Eur Phys J C 71:1547. https://doi.org/10.1140/epjc/s10052-011-1547-z, arXiv:1009.2450
126. Alioli S, Nason P, Oleari C, Re E (2011) Vector boson plus one jet production in Powheg. JHEP 01:095. https://doi.org/10.1007/JHEP01(2011)095, arXiv:1009.5594

Chapter 3
The LHC and the CMS Detector

This Chapter describes the main features of the Large Hadron Collider (LHC), the machine that provides the high energy collisions used to study the interactions between fundamental particles, and of the Compact Muon Solenoid (CMS) detector, which is used to record the collision data used in this thesis. The Chapter is structured as follows: details about the LHC are given in Sect. 3.1, while the CMS detector is described in Sect. 3.2. The CMS trigger system is then described in Sect. 3.3, and the measurement of the integrated luminosity delivered by the LHC, which is of particular importance for the work presented in this thesis, is discussed in Sect. 3.4.

3.1 The Large Hadron Collider

The Large Hadron Collider [1] is a circular proton-proton collider operating at the CERN laboratories in Geneva, designed to operate at the centre-of-mass energy of 14 TeV. The ring, installed in the 27 km tunnel previously used for the Large Electron Positron (LEP) collider [2], is used to generate two proton beams circulating in opposite directions. A schematic representation of the CERN accelerator complex is given in Fig. 3.1. The proton bunches are accelerated in steps using a linear accelerator, boosters, and synchrotrons, and are injected in the LHC with an energy of 540 GeV, where they are further accelerated by means of a system of superconducting magnets. Collisions between proton bunches are achieved every 25 ns at the four interaction points where the main experiments (ALICE [3], ATLAS [4], CMS [5], LHCb [6]) are located (Fig. 3.1). The LHC can also be used to produce collisions between heavy ions or between protons and heavy ions, which are however not used in this thesis.

The machine-dependent parameters of interest for a physics analysis are the centre-of-mass energy of the collisions, the instantaneous and integrated luminosity, and the number of simultaneous collisions within the same bunch crossing. The

M. M. Defranchis, *First Measurement of the Running of the Top Quark Mass*, Springer Theses, https://doi.org/10.1007/978-3-030-90376-3_3

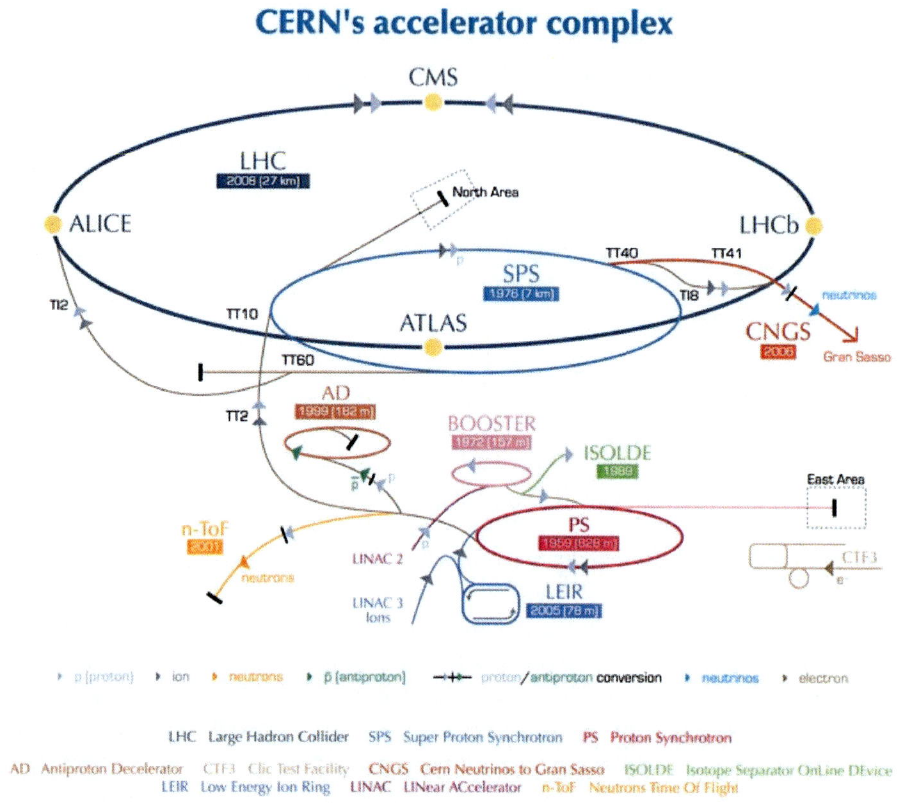

Fig. 3.1 The accelerator complex at the CERN laboratories in Geneva [7]

instantaneous luminosity $d\mathcal{L}/dt$ is a time-dependent quantity that relates the cross section σ of a given process to its production rate through the relation

$$\frac{dN}{dt} = \frac{d\mathcal{L}}{dt}\sigma, \tag{3.1}$$

The luminosity depends on the beam orbit frequency, the number of bunches in each beam, the number of protons per bunch, and the beam profile area.

The integral over time of the instantaneous luminosity is referred to as integrated luminosity, and relates the total cross section of a process to the total number of events N of a given process:

$$\mathcal{L} = \int \frac{d\mathcal{L}}{dt}\,dt = \frac{N}{\sigma}. \tag{3.2}$$

The LHC was designed to achieve an instantaneous luminosity of $10^{34}\mathrm{cm}^{-2}\mathrm{s}^{-1}$. So far, the LHC has completed two successful runs of data taking: Run1 (2010–2012), at the centre-of-mass energies of 7 and 8 TeV, and Run2 (2015–2018), at the centre-of-

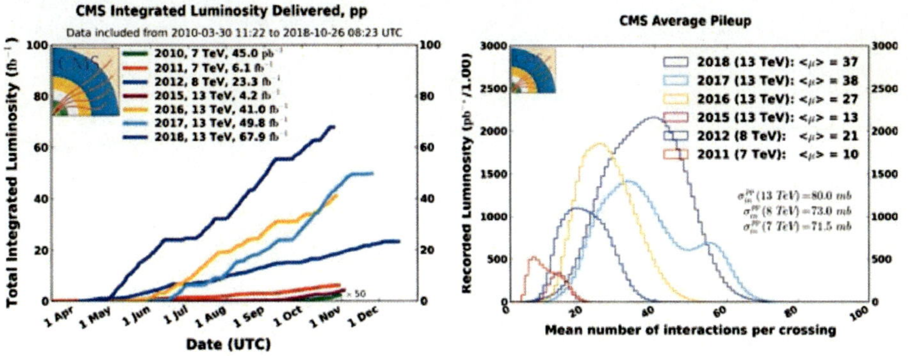

Fig. 3.2 Integrated luminosity collected by the CMS experiment (left) and average number of simultaneous pp interactions per bunch crossing (right) by year of data taking [8]

mass energy of 13 TeV. Special low-luminosity runs have also been conducted at the $\sqrt{s} = 2.67$ and 5.02 TeV. A total integrated luminosity of 192.3 fb^{-1} was delivered to CMS by the LHC so far, 162.9 fb^{-1} of which at $\sqrt{s} = 13$ TeV. Furthermore, a peak luminosity of above $2 \cdot 10^{34}\text{cm}^{-2}\text{s}^{-1}$ was achieved during Run2, which amounts to more than twice the design value. The integrated luminosity collected by the CMS experiment by year of data taking is shown in Fig. 3.2 together with the average number of simultaneous proton-proton (pp) interactions per bunch crossing, also referred to as pileup. The main method used to measure the integrated luminosity is described in Sect. 3.4.

3.2 The Compact Muon Solenoid

The Compact Muon Solenoid [5] is a multi-purpose experiment operating at the CERN LHC, designed according to the concept of cylindrical layers around the beam axis. The central feature of the CMS apparatus is a superconducting solenoid of 6 m of internal diameter which provides a magnetic field of 3.8 T. The CMS detector consists of several subdetector layers, as illustrated in Fig. 3.3. A silicon pixel and strip tracker, a lead tungstate crystal electromagnetic calorimeter (ECAL), and a brass and scintillator hadron calorimeter (HCAL) are located within the volume of the solenoid, while muons are detected in gas-ionization chambers embedded in the steel flux-return yoke outside the solenoid. Each subdetector consists of a central part of cylindrical geometry, referred to as the barrel, and of endcap disks perpendicular to the beam axis. Additionally, forward calorimeters extend the pseudorapidity coverage provided by the barrel and endcap detectors.

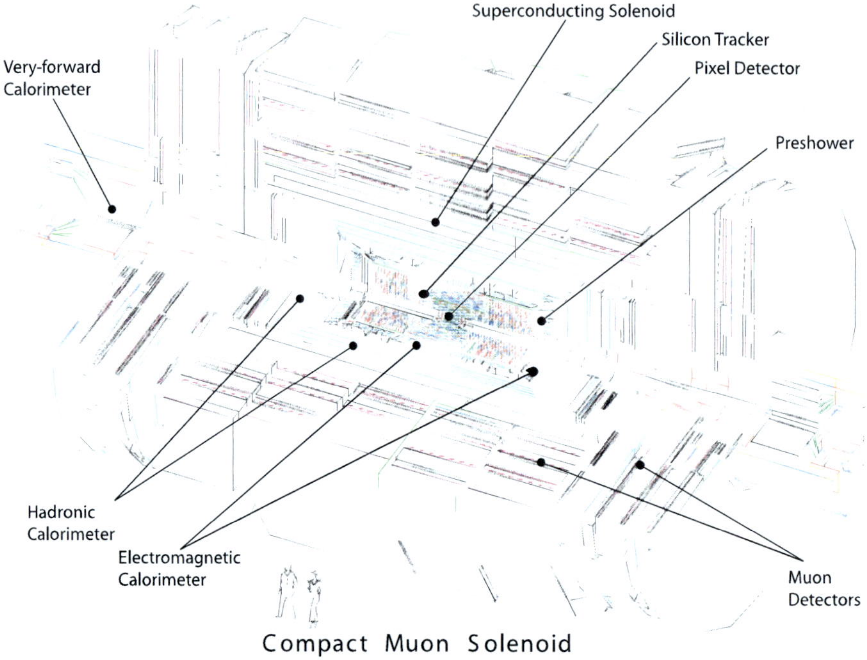

Fig. 3.3 A schematic representation of the CMS detector and its components [5]

3.2.1 The Coordinate System

The coordinate system adopted by CMS has its origin centred at the nominal collision point inside the detector, the y-axis pointing vertically upward, the x-axis pointing radially inward towards the centre of the LHC, and the z-axis along the direction of the beam. The azimuthal angle ϕ is measured from the x-axis in the $x - y$ plane and the polar angle θ is measured from the z-axis. The distance from the interaction point is defined as $r = \sqrt{x^2 + y^2}$ and the pseudorapidity is defined as $\eta = -\ln\tan(\theta/2)$. The momentum and energy transverse to the beam direction, denoted with $p_T$and E_T respectively, are computed from the x and y components. The imbalance of energy measured in the transverse direction is referred to as missing transverse energy, or missing E_T, and is denoted with E_T^{miss} or MET.

3.2.2 The Tracking System

The CMS tracking system is designed to provide efficient and precise measurements of the trajectories of charged particles and to precisely reconstruct secondary vertices. This latter aspect is of particular importance in the context of heavy-flavour tagging, as discussed in Chap. 4. In addition, high granularity and fast response are required

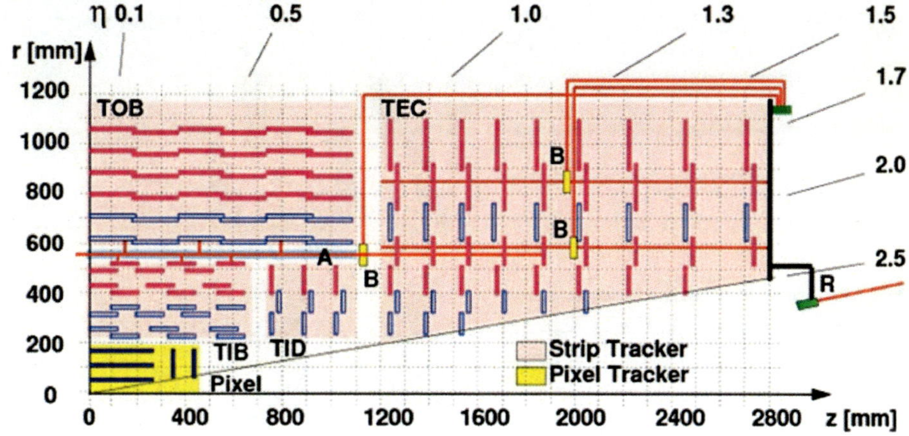

Fig. 3.4 A quarter of the CMS silicon tracker in the $r - z$ view. The strip tracker consists of several parts: the tracker inner barrel (TIB), outer barrel (TOB), inner disks (TID) and endcaps (TED) [9]

in order to reliably associate each track to the correct bunch crossing and interaction vertex.

The CMS tracker is composed of a pixel detector and a silicon strip detector (Fig. 3.4). The pixel detector consists of three barrel layers (BPIX) and two endcaps in the forward region (FPIX), while the strip tracker comprises several parts: the tracker inner and outer barrels (TIB and TOB), consisting of four and six concentric layers respectively; the tracker inner disks (TID), with three disks on each side; and the tracker endcap (TEC), with nine disks on each side, extending the coverage of the tracker up to $|\eta| = 2.5$. The tracking system comprises a total of 66 million pixels and 10 million strip sensors.

An upgraded version of the pixel detector with one additional pixel layer in the barrel region and one additional endcap disk per side was installed during the LHC shutdown between the 2016 and 2017 data takings [10]. In the upgraded system, the innermost layer and disks are located closer to the beam pipe in order to reconstruct the position of the interaction vertices with improved precision. A comparison between the original and upgraded versions of the pixel detector can be found in Fig. 3.5.

3.2.3 *The Electromagnetic Calorimeter*

The CMS ECAL is a hermetic and homogeneous calorimeter made of lead tungstate ($PbWO_4$) crystals. With an overall depth of about 25 radiation lengths, it is capable of absorbing more than 98% of the energy deposited by electrons and photons with energies up to 1 TeV.

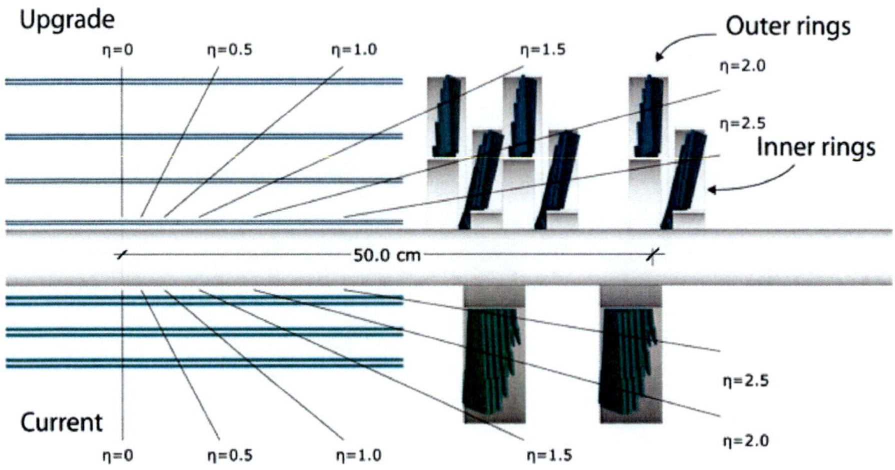

Fig. 3.5 Comparison between the original (indicated with "Current") and the upgraded versions of the CMS pixel detector [10]

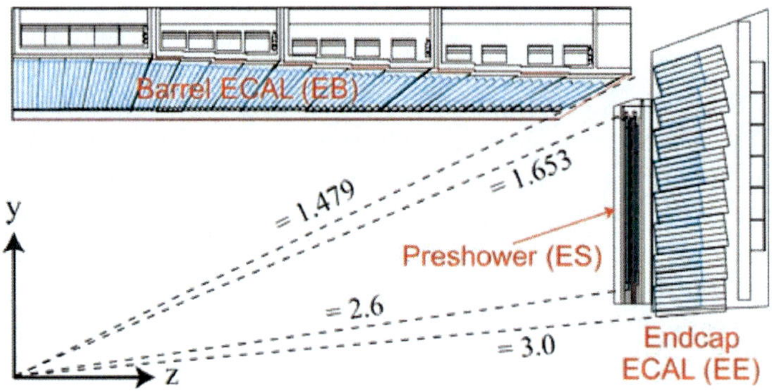

Fig. 3.6 Schematic view of the CMS electromagnetic calorimeter in the $r - z$ plane, divided into its subcomponents: ECAL barrel (EB), endcap (EE) and preshower (ES) [12]

A schematic representation of the ECAL is given in Fig. 3.6. Modules of crystals are mounted in the barrel ($|\eta| < 1.479$) and endcap ($1.635 < |\eta| < 3.0$) regions, and a preshower detector is placed before the endcap crystals. In the barrel section, an energy resolution of about 1% is achieved for unconverted or late-converting photons that have energies of the order of a few tens of GeV. The energy resolution for the remaining barrel photons is about 1.3% up to a pseudorapidity of $|\eta| = 1$, and rises to about 2.5% at $|\eta| = 1.4$. In the endcaps, the energy resolution for unconverted or late-converting photons is about 2.5%, while for the remaining endcap photons is between 3 and 4% [11].

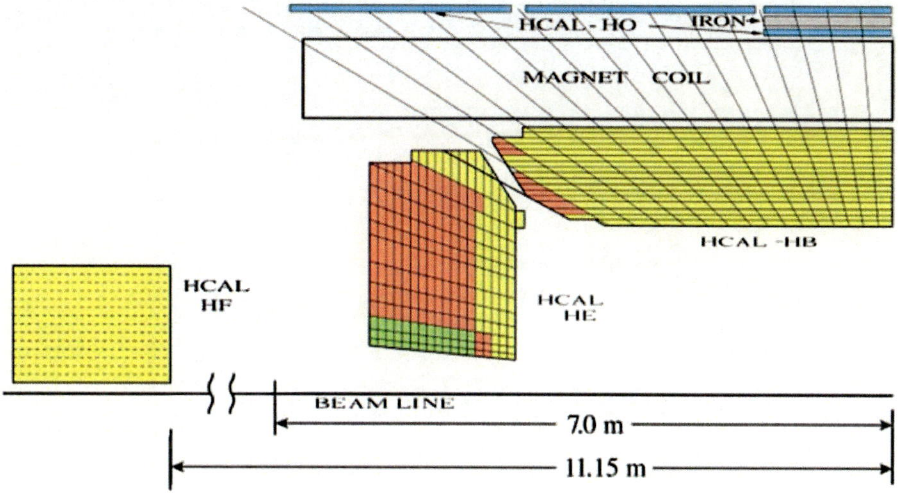

Fig. 3.7 Schematic view of the CMS hadron calorimeter in the $r - z$ plane, divided into its subcomponents: HCAL barrel (HB), endcap (HE), outer (HO) and forward (HF) [13]

3.2.4 *The Hadron Calorimeter*

The CMS HCAL consists of a set of sampling calorimeters, as illustrated in Fig. 3.7: the HCAL barrel (HB), covering the pseudorapidity range $|\eta| < 1.3$; the HCAL endcap (HE), in the range $1.3 < |\eta| < 3$; the outer calorimeter (HO); and the forward calorimeter (HF), that extends the coverage up to $|\eta| = 5.2$.

The barrel and endcap parts consist of alternating layers of brass absorbers and plastic scintillators, while the outer calorimeter is placed outside the central solenoid and utilizes the CMS magnet coil and the steel of the magnet return yoke as absorbers. The forward calorimeter is designed to be resistant to the intensive radiation conditions in the forward regions, and it uses steel as an absorber and quartz fibres as sensitive material.

3.2.5 *The Muon System*

The muon system is designed to efficiently detect muons and to determine their trajectories. It is located outside the solenoid and covers the pseudorapidity region $|\eta| < 2.4$. It is composed of three different types of gaseous detectors embedded in the flux-return yoke of the magnetic field, as illustrated in Fig. 3.8. Drift tube stations (DTs) and cathode strip chambers (CSCs) are located in the barrel and endcap regions, respectively, while and resistive plate chambers (RPCs) are located in both regions. The RPCs are characterized by responses of the order of 1 ns, and are therefore

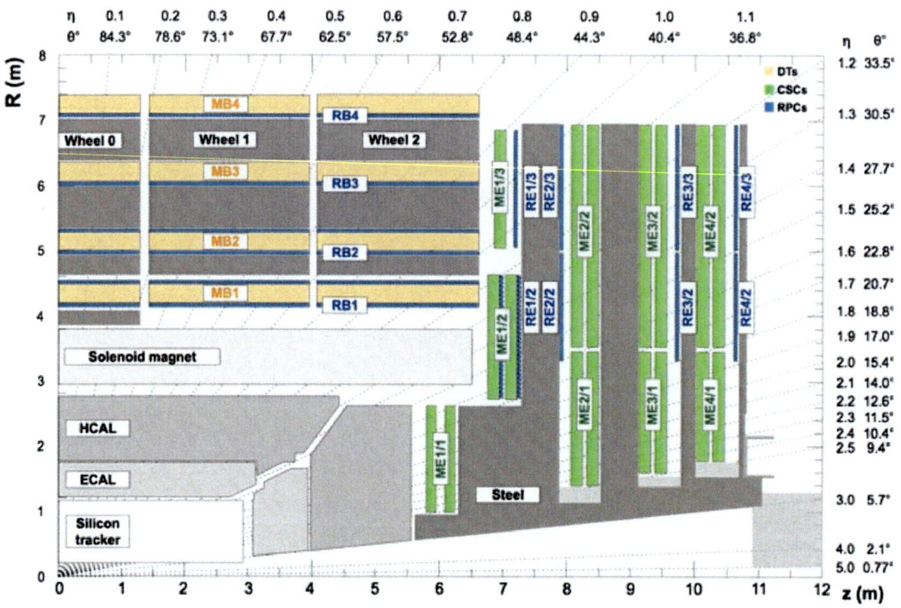

Fig. 3.8 Layout of one quadrant of CMS. The drift tube stations are labelled MB (muon barrel) and the cathode strip chambers are labelled ME (muon endcap). Resistive plate chambers are mounted in both the barrel and endcaps of CMS, where they are labelled RB and RE, respectively [14]

mainly used for the purpose of the trigger system (Sect. 3.3). The muon system can achieve a spacial resolution of about 100 μm and a momentum resolution better than 2% for muons with $p_T$up to about 100 GeV [14].

3.3 The CMS Trigger System

At the LHC, collisions between proton bunches are achieved every 25 ns, corresponding to a bunch crossing frequency of 40 MHz. However, only a small fraction of the events are of potential interest for the physics program of CMS, and the capability to store collision data is limited by the rate at which they can be saved on disk for offline analysis.

The CMS trigger system is designed to select potentially interesting events by applying a sequence of selection criteria. The system is implemented in two separate levels: the first one (L1), implemented in custom hardware, is designed to reduce the output rate to about 100 kHz by means of an approximate event reconstruction and loose selection criteria. Events selected by the L1 trigger are further processed by the high-level trigger (HLT), a software farm that refines the event reconstruction selection, reducing the output rate to a few kHz. Events selected by the HLT are saved for offline analysis.

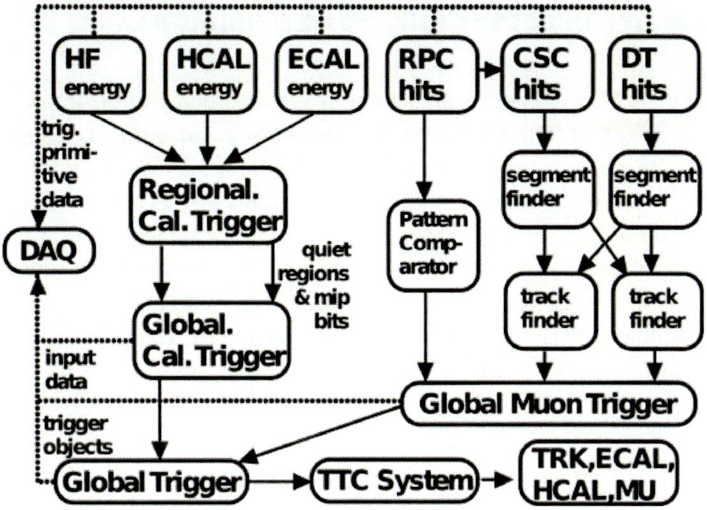

Fig. 3.9 Overview of the structure of the CMS L1 trigger system [16]

The L1 trigger is a system that combines information from the calorimeters and the muon system in order to reconstruct candidate objects and select events. The system has a fixed latency of 4 μs, after which an event is either accepted or rejected. The system comprises a calorimeter and a muon trigger, as illustrated in Fig. 3.9. The calorimeter trigger makes use of information from all calorimeters to reconstruct candidate jets, electrons or photons, and τ leptons, as well as the missing E_{T} in the event. Similarly, the global muon trigger combines the hit information in RPCs, CSCs, and DTs to reconstruct muon candidates. Finally, a global trigger applies selection criteria to the reconstructed objects, and selected events are passed to the HLT. Before the beginning of Run2, an upgraded version of the L1 trigger system was installed. In the upgraded trigger, additional interconnections between the calorimeter and the muon systems are provided, which allow to implement algorithms such as muon isolation or muon-jet distance [15].

The event reconstruction performed at the HLT is mainly based on the algorithms used for offline analysis (Chap. 4), adapted to match the timing constraints imposed by the software farm. Both the L1 and the HLT output rates can be adjusted by prescaling the number of selected events passing specific selection criteria. A complete sequence of L1 and HLT selection criteria, including any prescale, is referred to as a trigger path. Unlike in the case of offline analysis, the trigger selection is a non-reversible process, and discarded events cannot be recovered.

3.4 Measurement of the Integrated Luminosity

The determination of the integrated luminosity recorded by the CMS detector is particularly important in the context of cross section measurements, as explained in Chap. 5. One of the most precise methods to measure the integrated luminosity is the so-called pixel cluster counting (PCC) [17, 18]. Pixel clusters are group of pixel cells formed by combining side-by-side and corner-by-corner adjacent pixel cells in which a signal was detected. The method is based on the assumption that the number of reconstructed pixel clusters is linearly related to the number of interactions per bunch crossing, and hence to the instantaneous luminosity. The assumption is justified by the fact that the probability for one of the 66 million pixels to be hit by two different tracks from the same bunch crossing is negligible.

The instantaneous luminosity per colliding bunch can be expressed in terms of the total inelastic pp interaction cross section σ_0, the LHC orbit frequency $f = 11.25$ kHz, and the average number of simultaneous pp interactions per bunch crossing μ as:

$$\frac{1}{N_b}\frac{\mathrm{d}\mathcal{L}}{\mathrm{d}t} = \frac{\mu f}{\sigma_0}, \tag{3.3}$$

where N_b is the number of proton bunches in the beam. By defining $\langle N_{\text{cluster}}\rangle$ as the average number of pixel clusters per bunch crossing and σ_{pixel} as the product between σ_0 and the average number of clusters per pp interaction, the equation above can be written as:

$$\frac{1}{N_b}\frac{\mathrm{d}\mathcal{L}}{\mathrm{d}t} = \frac{\langle N_{\text{cluster}}\rangle f}{\sigma_{\text{pixel}}}. \tag{3.4}$$

The remaining unknown quantity σ_{pixel} can be determined by means of Van der Meer (VdM) scans [19]. The VdM technique consists in measuring the relative event rate as a function of the beam separation along the transverse axes, Δx and Δy. Assuming a Gaussian beam density in the transverse plane, the instantaneous luminosity per colliding bunch can be expressed as a function of the displacement as:

$$\frac{1}{N_b}\frac{\mathrm{d}\mathcal{L}}{\mathrm{d}t}(\Delta x, \Delta y) = \frac{f N_1 N_2}{2\pi \Sigma_x \Sigma_y} \exp\left(-\frac{\Delta x^2}{2\Sigma_x^2} - \frac{\Delta y^2}{2\Sigma_y^2}\right), \tag{3.5}$$

where N_1 and N_2 denote the number of protons in each bunch, and Σ_x and Σ_y represent the beam spread along the x and y axes respectively. VdM scans are used to measure the quantities Σ_x and Σ_y, which are then used to determine σ_{pixel} using the relation

$$\sigma_{\text{pixel}} = \frac{2\pi \Sigma_x \Sigma_y}{N_1 N_2} \langle N_{\text{cluster}}^{\Delta=0}\rangle, \tag{3.6}$$

where $\langle N_{\text{cluster}}^{\Delta=0}\rangle$ is the average number of pixel clusters per bunch crossing at $\Delta x = \Delta y = 0$. With the knowledge of σ_{pixel}, Eq. 3.4 directly relates the average number of pixel clusters to the instantaneous luminosity. The integrated luminosity is then obtained by integrating the instantaneous luminosity over time. The total integrated luminosity collected by the CMS detector during 2016 data taking, which corresponds to the data set used in this thesis, is estimated to be 35.9 ± 2.5 fb^{-1} [20].

References

1. Brüning OS, Collier P, Lebrun P, Myers S, Ostojic R, Poole J, et al (2004) LHC design report, CERN Yellow reports: monographs. CERN, Geneva. https://doi.org/10.5170/CERN-2004-003-V-1
2. LEP design report (1984) CERN, Geneva. https://cds.cern.ch/record/102083
3. ALICE Collaboration (2008) The ALICE experiment at the CERN LHC, JINST 3:S08002. https://doi.org/10.1088/1748-0221/3/08/S08002
4. ATLAS Collaboration (2008) The ATLAS experiment at the CERN Large Hadron Collider, JINST 3:S08003. https://doi.org/10.1088/1748-0221/3/08/S08003
5. Collaboration CMS (2008) The CMS experiment at the CERN LHC. JINST 3:S08004. https://doi.org/10.1088/1748-0221/3/08/S08004
6. LHCb Collaboration (2008) The LHCb detector at the LHC. JINST 3:S08005. https://doi.org/10.1088/1748-0221/3/08/S08005
7. Lefèvre C The CERN accelerator complex. Complexe des accélérateurs du CERN. CERN-DI-0812015. https://cds.cern.ch/record/1260465
8. CMS Collaboration (2019) Public CMS luminosity information. https://twiki.cern.ch/twiki/bin/view/CMSPublic/LumiPublicResults. Accessed: May 2019
9. CMS Collaboration (2010) Alignment of the CMS silicon tracker during commissioning with cosmic rays. JINST 5:T03009. https://doi.org/10.1088/1748-0221/5/03/T03009. arXiv:0910.2505
10. CMS Collaboration (2012) CMS technical design report for the pixel detector upgrade, Technical report CERN-LHCC-2012-016; CMS-TDR-11, CERN, Geneva. https://cds.cern.ch/record/1481838
11. CMS Collaboration (2015) Performance of photon reconstruction and identification with the CMS detector in proton-proton collisions at $\sqrt{s} = 8$TeV. JINST 10:P08010. https://doi.org/10.1088/1748-0221/10/08/P08010, arXiv:1502.02702
12. Collaboration CMS (2014) The CMS ECAL performance with examples. JINST 9:C02008. https://doi.org/10.1088/1748-0221/9/02/C02008
13. CMS Collaboration (2010) Performance of CMS hadron calorimeter timing and synchronization using test beam, cosmic ray, and LHC beam data. JINST 5:T03013. https://doi.org/10.1088/1748-0221/5/03/T03013, arXiv:0911.4877
14. CMS Collaboration (2018) Performance of the CMS muon detector and muon reconstruction with proton-proton collisions at $\sqrt{s} = 13$ TeV. JINST 13:P06015. https://doi.org/10.1088/1748-0221/13/06/P06015, arXiv:1804.04528
15. CMS Collaboration (2013) CMS technical design report for the level-1 trigger upgrade, Technical report CERN-LHCC-2013-011. CMS-TDR-12, CERN, Geneva. https://cds.cern.ch/record/1556311
16. CMS Collaboration (2017) The CMS trigger system. JINST 12:P01020. https://doi.org/10.1088/1748-0221/12/01/P01020, arXiv:1609.02366
17. CMS Collaboration (2012) Absolute calibration of the luminosity measurement at CMS: winter 2012 update. CMS physics analysis summary CMS-PAS-SMP-12-008, CERN, Geneva. https://cds.cern.ch/record/1434360
18. CMS Collaboration (2013) CMS luminosity based on pixel cluster counting - summer 2013 update. CMS physics analysis summary CMS-PAS-LUM-13-001, CERN, Geneva. https://cds.cern.ch/record/1598864

19. van der Meer S (1968) Calibration of the effective beam height in the ISR, Technical report CERN-ISR-PO-68-31, CERN, Geneva
20. CMS Collaboration (2017) CMS luminosity measurements for the 2016 data taking period. CMS Physics Analysis Summary CMS-PAS-LUM-17-001, CERN, Geneva. http://cds.cern.ch/record/2257069

Chapter 4
Event Reconstruction in CMS

The reconstruction of an event implies the identification of all final state particles produced in a proton-proton interaction. Particles are identified based on their specific signatures in the CMS detector, as illustrated in Fig. 4.1, and their kinematic properties are measured by combining the information from the various detectors. The trajectories of charged particles in the magnetic field of the CMS apparatus are reconstructed from the hits in the tracking detector, and the curvature of the reconstructed tracks is used to estimate the particle momenta. Electrons and charged hadrons are then absorbed in the material of the ECAL and HCAL, respectively, and their energy is estimated based on the calorimetric deposits, while muons are detected in the muon system before escaping the CMS detector. Photons and neutral hadrons are also absorbed in the material of the calorimeters, but do not generate hits in the tracking detector. Neutrinos instead do not interact with the material of the CMS detector, and escape undetected. However, missing transverse energy in an event can be an indication of the presence of one or more final state neutrinos. The reconstructed particles are then combined to form more complex physics objects that are directly used in physics analyses.

This chapter describes the concepts and the most relevant details of the CMS event reconstruction algorithms, with particular emphasis on the aspects relevant for this thesis. The reconstruction of interaction vertices and charged tracks is described in Sect. 4.1, and the particle-flow algorithm is presented in Sect. 4.2. The reconstruction of jets is then discussed in Sect. 4.3, followed by a description of the b jet identification algorithms in Sect. 4.4. Studies related to the performance of b jet identification at the high-level trigger, obtained with significant contribution from the author of this thesis, are illustrated in Sect. 4.5 and documented in Ref. [1]. Finally, Sect. 4.6 describes the commissioning of a boosted $H \rightarrow b\bar{b}$ reconstruction algorithm, performed in the scope of this thesis and documented in Ref. [2].

M. M. Defranchis, *First Measurement of the Running of the Top Quark Mass*, Springer Theses, https://doi.org/10.1007/978-3-030-90376-3_4

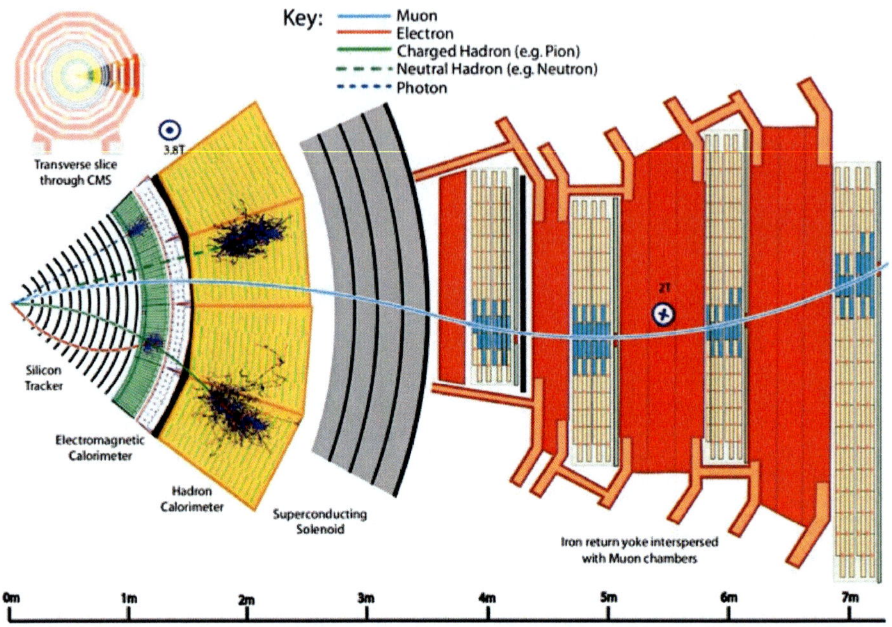

Fig. 4.1 Slice of the CMS detector and specific interactions of final-state particles with the different subdetectors. In this sketch, the muon and the charged hadron are positively charged, while the electron is negatively charged [3]

4.1 Reconstruction of Tracks and Vertices

At every bunch crossing, the collisions between protons give rise to interaction vertices which spread along the beam axis around the nominal interaction point, which can be reconstructed by combining information from reconstructed tracks, as illustrated in Fig. 4.2. Tracks are then associated to vertices, and the primary vertex (PV) is defined as the one with the largest value of summed physics-object p_T^2. The other vertices are referred to as pileup vertices. Tracks and vertices are also an essential input to the particle-flow algorithm (Sect. 4.2) and are crucial information in the identification (tagging) of heavy-flavour jets (Sect. 4.4).

In CMS, tracks are reconstructed from hits of charged particles in the pixel and strip tracker. The tracking algorithms are designed to optimize the track-finding efficiency while keeping under control the fraction of fake tracks, i.e. those formed from unrelated hits or including spurious hits. Hits are reconstructed in two steps: the first one, referred to as local reconstruction, consists in the clustering of signals above given thresholds in the tracker sensors. Pixel clusters are formed and projected onto orthogonal axes in a local coordinate system by summing the charge collected in pixels with the same coordinate along one of the axes. A first estimate of the hit position is then derived based on the geometry of the sensors and taking into account the Lorentz drift of the collected charge in the magnetic field. A more

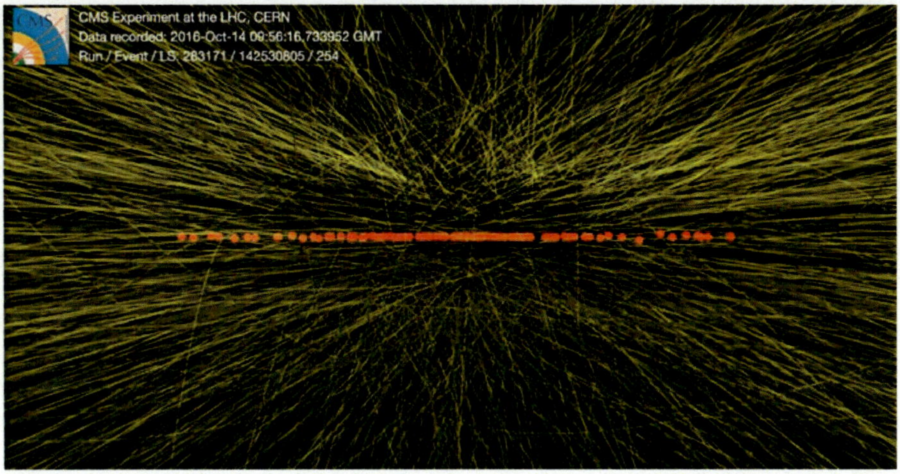

Fig. 4.2 A collision recorded by the CMS detector during the 2016 data taking [4]. The lines correspond to the reconstructed tracks, while the dots represent the reconstructed interaction vertices

sophisticated hit reconstruction follows, based on templates that take into account the irradiation status of the sensors. Similarly, strip clusters are formed combining sensors where a significant signal is detected. The hit position is then determined by the charge-weighted average of its strip positions, corrected for the Lorentz drift. The hit efficiency is found to be above 99.5%, both for pixel and tracker hits, excluding non-operative modules [5]. However, the efficiency depends on the instantaneous luminosity and on the trigger output rate, especially in the innermost layer of the pixel detector where the occupancy is higher. The resolution in the hit position is estimated to be between 20 (10) and 50 μm for the pixel (strip) detector, depending on the angle of incidence of the particle and on the width of the clusters [5].

In order to reconstruct tracks, the estimated hit positions are translated to a global coordinate system. In the conversion, the discrepancy between the assumed and the actual position of the sensors, which can be significantly larger than the hit resolution, has to be taken into account. The process of estimating the sensor positions is known as tracker alignment, and consists in the minimization of the residuals between hit position and reconstructed tracks. The alignment is performed with using cosmic-ray data and collision data. The final uncertainty in the alignment procedure is of the order of 1 μm in the pixel detector, and of the order of 10 μm in the strip detector [6].

Other important inputs to the track reconstruction are the LHC beam spot, i.e. the 3-dimensional profile of the LHC luminous region, and an approximate position of interaction vertices. These can be estimated using pixel tracks, i.e. tracks reconstructed using only the information from pixel hits. Since pixel tracking requires considerably shorter computing time than the full tracking, this technique is particularly important in the context of b-tagging at the HLT, as discussed in Sect. 4.5.

The CMS tracking algorithm, called Combinatorial Track Finder (CTF), is based on the concept of iterative tracking. At each iteration, track seeds are formed using a limited number of pixel hits, and an initial estimate of the track parameters is derived. A track-finding algorithm follows, based on the method of Kalman filter [7]. The filter uses track seeds to extrapolate the trajectory of the track to subsequent detector layers, taking into account the effect of multiple Coulomb scattering. At each layer, compatible hits are added to the trajectory, the track parameters are recalculated, and the resulting trajectory is extrapolated to the next layer. Finally, tracks that do not satisfy goodness-of-fit criteria are discarded. Tracks that are easier to reconstruct, e.g. those with large p_T (and therefore less pronounced curvature) and produced close to the interaction point, are reconstructed first. Hits associated to these tracks are then removed, reducing the complexity of subsequent iterations. A maximum of 6 iterations is performed, each time with decreasingly stringent requirements on the track seeds. In particular, the last iterations are optimized to reconstruct tracks with lower p_T or with larger displacement from the interaction vertex. The efficiency of reconstructing tracks with $p_T > 1$ GeV is found to be larger than 99% for isolated muons over the entire coverage of the tracker, while efficiencies for electrons and pions range between 80 and 99%, depending on the track pseudorapidity. Fake rates are at the level of 5% in the barrel and of up to 15% in the endcap and transition regions. The resolution on the track p_T significantly depends on the p_T and η of the tracks, and is below 1% for central muons with p_T between 1 and 10 GeV [5].

Interaction vertices are reconstructed using available tracks. High-quality tracks compatible with originating in the beam spot are clustered on the basis of their coordinates along the z-axis and of the distance of closest approach to the centre of the beam spot. The coordinates of the vertices are then estimated using an adaptive vertex fitter [8]. Based on the coordinates of the tracks and vertices, a probability for a track to belong to a particular vertex is calculated and assigned to a track as a weight between 0 and 1. Vertex weights are then derived as the sum of the weights of all associated tracks, and interaction vertices with weights below predefined thresholds are discarded. The vertex reconstruction efficiency is found to be close to 100% for vertices with more than two tracks, and is of about 98% for vertices with only two tracks [5]. The resolution in the vertex coordinates strongly depends on the number of tracks associated to the vertex and on the average p_T of the tracks, ranging between 10 and 100 μm [5].

4.2 The Particle-Flow Algorithm

The particle-flow (PF) algorithm [3] is designed to provide a global event description by combining information from different subdetectors. The output of the PF consists of a list of candidates classified as electrons, photons, muons, charged hadrons, or neutral hadrons, as illustrated in Fig. 4.3. The algorithm improves the performance in the identification of leptons, in the determination of the missing transverse momentum, and in the identification of pileup tracks. The angular and momentum resolution

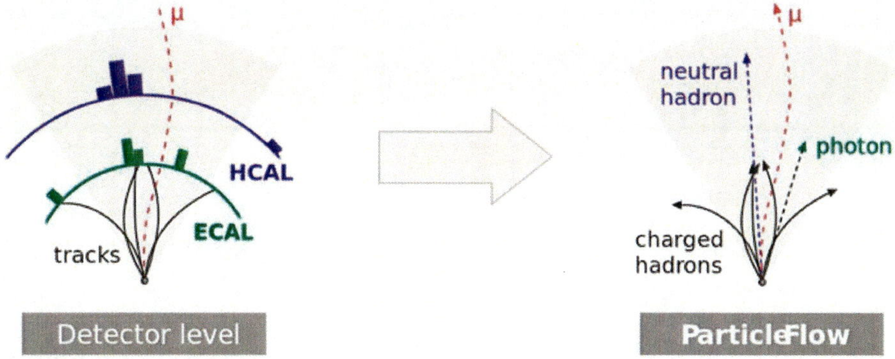

Fig. 4.3 The particle-flow algorithm combines information from various subdetectors to provide a global description of the event in terms of electrons, photons, muons, charged hadrons, and neutral hadrons [9]

of jets is also significantly improved when combining information from the tracker and the calorimeters [3].

The algorithm consists of two separate steps: the reconstruction of the PF elements, and the link between the reconstructed elements. Particle-flow elements include charged tracks, muon tracks, and calorimeter clusters. Muon tracks are fitted by combining information from the muon chambers and from the tracking system, while calorimeter clusters are formed using a dedicated reconstruction procedure that combines information from neighbouring calorimeter cells [3]. The energy and position of the cluster are then estimated by means of a likelihood fit. In the final step, tracks other than muon tracks are linked to calorimetric deposits in order to create candidate electrons, photons, charged hadrons, and neutral hadrons.

4.3 Jet Clustering and Calibration

In CMS, jets are reconstructed using a sequential recombination algorithm known as anti-k_T [10]. The algorithm is designed to be safe against infrared and collinear singularities, i.e. insensitive to soft and collinear gluon emission, and collinear gluon splitting. Particle-flow candidates are sequentially recombined following the procedure described below. A distance d_{ij} between entities (i.e. particles or combinations of particles) i and j is introduced together with a distance d_{iB} between entity i and the beam. These two quantities are defined, respectively, as:

$$d_{ij} = \min({p_T}_i^{-2}, {p_T}_j^{-2})\frac{\Delta R_{ij}^2}{R^2}\,, \tag{4.1}$$

$$d_{iB} = {p_T}_i^{-2}\,, \tag{4.2}$$

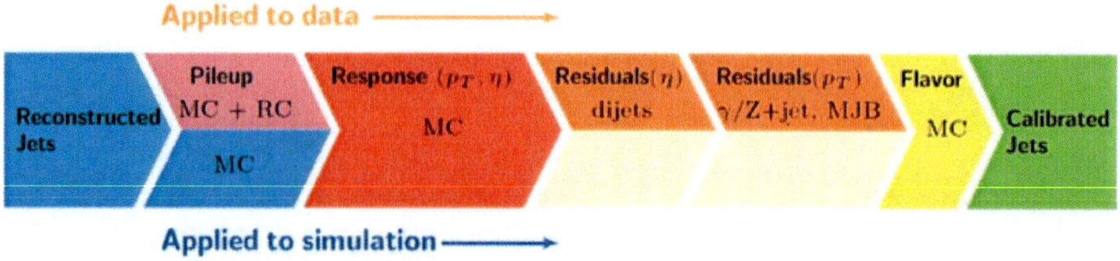

Fig. 4.4 Representation of the jet energy calibration procedure in data and MC [16]

where $\Delta R^2_{ij} = (\Delta\eta_{ij})^2 + (\Delta\phi_{ij})^2$, and p_{Ti} is the transverse momentum of entity i. The distance parameter R controls the size of the jet and can be chosen arbitrarily. In CMS, jets are reconstructed using $R = 0.4$. The quantities d_{ij} and d_{iB} are calculated for all possible combinations of entities in the event. If the smallest d_{ij} is smaller than the smallest d_{iB}, entities i and j are combined to form a new entity. Otherwise, entity i is removed and called a jet. The procedure is iterated until no particles are left. The jet momentum is then defined as the vectorial sum of the momenta of all particles in the jet. In order to mitigate the effect of pileup interactions, charged particles associated to pileup vertices are removed prior to the jet clustering. This procedure is knows as charged hadron subtraction [11].

The energy of the jets is calibrated using a factorized approach where subsequent corrections that account for different effects are applied, as illustrated in Fig. 4.4. First, the contribution from neutral pileup is subtracted based on the average p_T per unit area expected in pileup jets [12–14]. The pileup offset is determined from simulation, and a correction is derived from data [15]. Corrections to the simulated jet response, i.e. the ratio between the momentum of a reconstructed jet and the corresponding generated jet, are then derived as a function of the jet p_T and η and applied to both data and simulation. Here, generated jets are defined at the particle level after hadronization, and neutrinos are excluded from the jet clustering. These corrections are derived separately for jets initiated by quarks of different flavours and for gluon jets. Finally, residual differences in the jet response in data and simulation are estimated using Z+jets, multijet, and dijet events [12], and correction factors are applied to the simulation. Similarly, the jet p_T resolution in the simulation is corrected to match the one in the data, using corrections derived from dijet and γ+jet events based on momentum conservation [12].

4.4 Identification of Jets Originating from b Quarks

The identification of jets originating from b quarks, also called b jets, is extremely useful in physics analyses involving top quarks to discriminate the signal against background processes. These techniques, commonly referred to as b tagging, also played a crucial role in the recent observation of the coupling of the Higgs boson

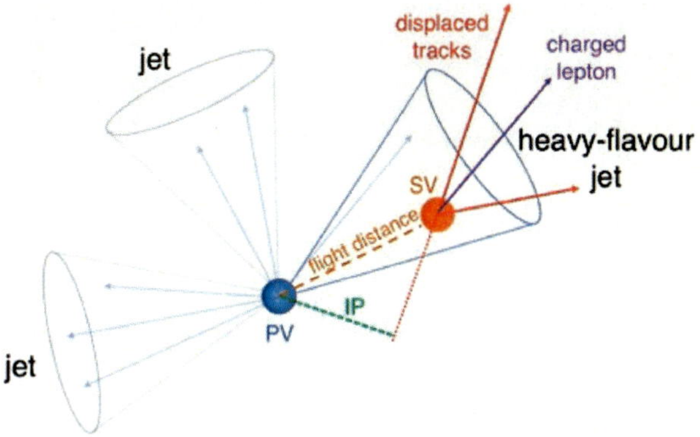

Fig. 4.5 Schematic representation of the properties of a b jet: b hadrons decay after propagating from the primary vertex for up to a few cm, giving rise to secondary vertices and displaced tracks [2]. In about 20% of the cases, the b hadron decay products include a charged lepton

to b quarks [17, 18] and are often used in searches for physics beyond the standard model.

Jets originating from b quarks contain b hadrons, which have a lifetime of about 1.5 ps and mass of about 5 GeV. For this reason, b hadrons produced with momenta sufficiently larger than their mass can propagate from the PV for up to a few cm before decaying. This gives rise to secondary vertices (SVs) containing tracks displaced from the PV, as illustrated in Fig. 4.5. Such signatures are exploited by b tagging algorithms in order to construct variables that provide optimal separation between b jets and light jets, i.e. jets initiated by light (u,d,s) quarks or gluons. Jets originating from c quarks have intermediate properties between light jets and b jets, and can be identified using dedicated c tagging techniques [2].

One of the most important variables in this context is the track impact parameter (IP), defined as the distance of closest approach of each track to the primary vertex (Fig. 4.5). Conventionally, a positive sign is assigned to the IP if the angle between the jet direction and the IP is smaller than $\pi/2$, while a negative sign is assigned otherwise. Tracks originating from SVs are expected to have mainly positive IP, while the IP values of tracks originating from the PV are expected to be symmetrically distributed around zero, due to detector resolution effects. In order to take the detector resolution into account, an additional variable referred to as IP significance (IP/σ) is introduced, defined as the ratio between the IP and its uncertainty. The IP and its significance can be defined either in 3 dimensions (3D) or in the plane transverse to the beam line (2D). Typical distributions of the 3D and 2D IP significances are shown in Fig. 4.6 for jets of different flavours.

Additional useful information to identify b jets can be accessed by reconstructing SVs. In CMS, SVs are reconstructed using the inclusive vertex finder (IVF) algorithm [19], which makes use of all available tracks in the event independently of the jet clustering. The IVF is seeded with tracks exceeding predefined values of the 2D IP significance and the 3D IP, and tracks are clustered on the basis of geometrical criteria. A first fit of secondary vertices is then performed, after which tracks compatible with the primary vertex are removed. A second fit with the remaining tracks

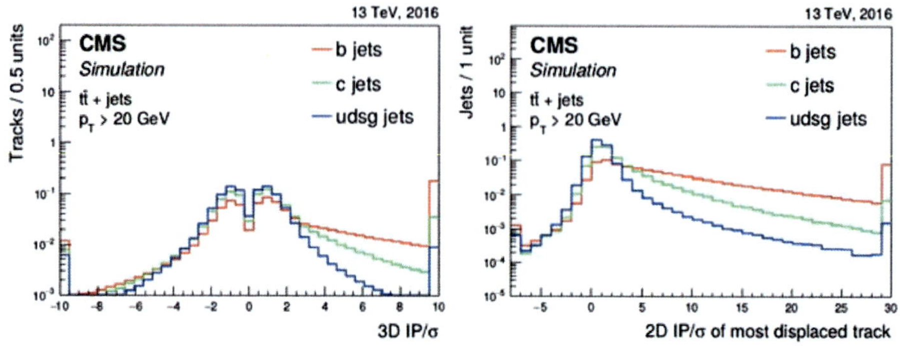

Fig. 4.6 Distribution of the 3D (left) and 2D (right) IP significance for jets of different flavours in $t\bar{t}$ events with additional jets [2]

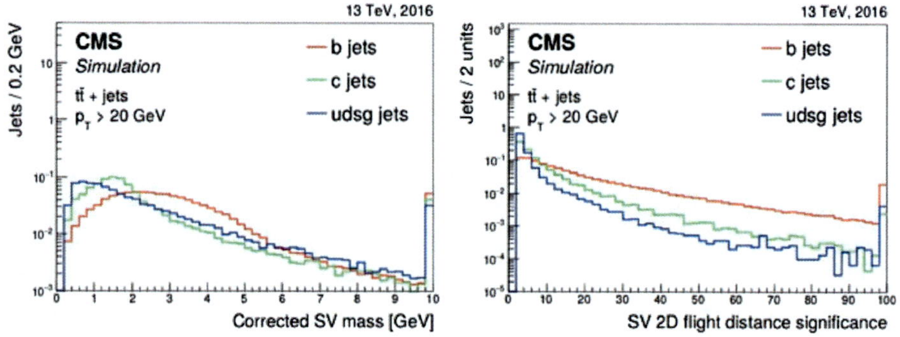

Fig. 4.7 Distribution of the corrected SV mass (left) and of the 2D FD significance (right) for jets of different flavours in $t\bar{t}$ events with additional jets [2]

follows, and selection criteria are applied to remove vertices that share a significant number of tracks.

The distance of a SV from the PV is referred to as flight distance (FD), and can be useful in order to identify b jets. As in the case of the IP, the FD can be defined in 3D or in the plane transverse to the beam direction (2D). Similarly, the FD significance is defined as the ratio between the FD and its uncertainty. Other useful discriminating variables include the number of reconstructed SVs and their corrected mass, which is an observable that estimates the mass of the decaying hadron by correcting for the effect of particles that were not detected or associated to the SV. The distribution of the corrected SV mass and of the 2D FD significance for jets of different flavours are shown in Fig. 4.7.

The variables described above are used as inputs to the combined secondary vertex (CSVv2) algorithm [2], a commonly used binary classifier based on a fully-connected neural network with a fixed amount of input variables and a single output node. Jets with $p_T > 20\,\text{GeV}$ and at least two high-quality tracks with $p_T > 1\,\text{GeV}$ are considered, and each jet is assigned a score between 0 and 1 that represents the probability of the jet being a b jet. Instead, a score of -1 is assigned to jet that do not

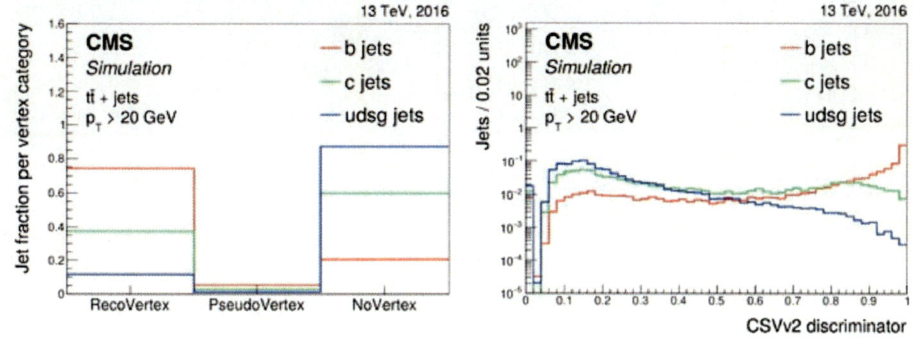

Fig. 4.8 Left: vertex categories used for the training of the CSVv2 tagger [2]. The pseudovertex category is defined by jets having two tracks with high IP significance, but no reconstructed SV. Right: output of the CSVv2 tagger for jets of different flavour [2]. Here, jets with negative score are added to the first bin

satisfy the track requirements described above. The algorithm is trained in categories of SV multiplicity, taking into account features of up to four tracks, ranked by 2D IP significance. The different categories are then combined using a likelihood ratio which takes into account the expected flavour composition in $t\bar{t}$ events, as show in Fig. 4.8 (left). The full list of variables considered in the training can be found in Ref. [2], and the output of the CSVv2 tagger for jets of different flavour is shown in Fig. 4.8 (right).

Operating points, or working points, are recommended b jet selection criteria for which data-to-simulation corrections are provided. Standard operating points for the CSVv2 tagger are defined based on the probability of a light jet to be wrongly identified (misidentified) as a b jet. In particular, a loose, a medium, and a tight working point are derived, corresponding to misidentification probabilities of about 10, 1, and 0.1%, respectively. The corresponding b tagging efficiencies are estimated to be about 81, 63, and 41%, respectively, while the misidentification probability for c jets are about 37, 12, and 2.2%, respectively. Data-to-simulation corrections to the b tagging efficiency are derived using multijet events enriched in jets containing soft muons and $t\bar{t}$ events. Since the probability of a b hadron to decay into a muon is about 20% [20], muon-enriched multijet events are expected to contain a large fraction of b jets. The measurements described in Chaps. 5 and 7 make use of the tight working point of the CSVv2 algorithm.

An improved version of the CSVv2 algorithm called DeepCSV [2] has also been developed. The DeepCSV tagger is based on a deep neural network architecture that further improves the performance of the classification [21]. In addition, unlike CSVv2, DeepCSV is a multi-classifier trained to distinguish between five exclusive jet flavour categories defined according to the jet content. The categories correspond to jets containing exactly one b hadron (b), exactly two b hadrons (bb), exactly one c hadron and no b hadrons (c), exactly two c hadrons and no b hadrons (cc), or none of the above (udsg). The probability for a jet to belong to the flavour category f is indicated with $P(\mathrm{f})$. The DeepCSV tagger uses the same input variables as CSVv2,

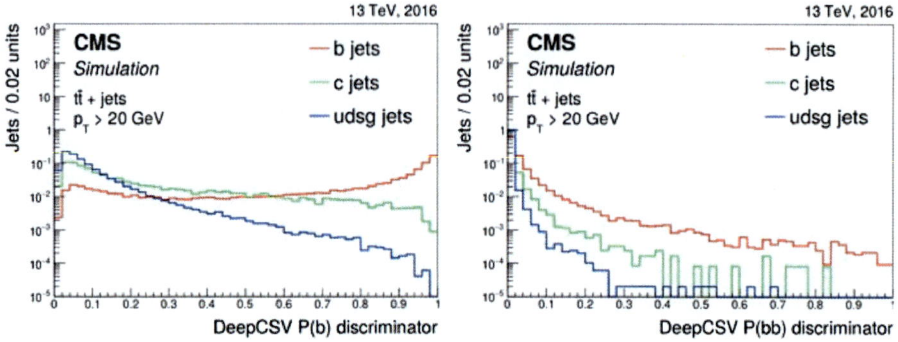

Fig. 4.9 Distribution of the DeepCSV outputs $P(\text{b})$ (left) and $P(\text{bb})$ (right) for jets of different flavours in $\text{t}\bar{\text{t}}$ events with additional jets [2]

with information of up to six tracks. Unlike in the case of CSVv2, the DeepCSV training is performed in all vertex categories simultaneously. It was demonstrated that the best b tagging performance with DeepCSV can be obtained by using the quantity $P(\text{b}) + P(\text{bb})$ as binary classifier [2]. The distributions of the $P(\text{b})$ and $P(\text{bb})$ outputs of the DeepCSV algorithm for jets of different flavours are shown in Fig. 4.9. Working points are defined as for CSVv2, and the corresponding b tagging efficiencies are estimated to be about 84, 68, and 50% for the loose, medium, and tight working point, respectively. A further improvement in the b tagging performance has been achieved starting from the 2017 data taking thanks to the upgraded tracking detector [22] and the newly-developed DeepJet tagger [23].

4.5 Identification of b Jets at the High Level Trigger

The b tagging algorithms described in Sect. 4.4, CSVv2 and DeepCSV, are also deployed at the HLT in order to select events containing b jet candidates. This is particularly important in the context of analyses that require the presence of b jets in fully-hadronic events, e.g. of $\text{t}\bar{\text{t}}$ measurements in which both W bosons are required to decay into quarks. On the other hand, analyses requiring the presence of leptons usually utilize leptonic triggers, which can achieve higher efficiency. In Run2, the CSVv2 algorithm has been successfully used to collect data during 2016 and 2017, and it was replaced with DeepCSV at the beginning of the 2018 data taking. The necessary studies and developments for the transition were performed using 2017 data, with significant contribution from the author of this thesis to the coordination of the effort.

At the HLT, two versions of each tagger are available: one, similar to the offline version of the tagger, uses track candidates reconstructed with the PF algorithm; the other, instead, utilizes tracks reconstructed using pixel hits which are then matched to jets clustered using information solely from the calorimeters (called Calo jets).

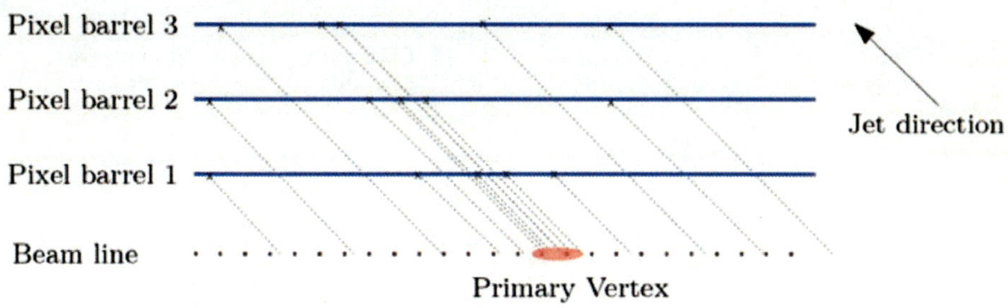

Fig. 4.10 Illustration of the fast primary vertex reconstruction used to determine the approximate position of the primary vertex along the beam line using pixel hits only [2]

The first version is referred to as PF b tagging, while the second one is called Calo b tagging. The advantage of PF b tagging lies in the more precise reconstruction of tracks and vertices, although at the cost of longer computing time, while Calo b tagging is based an approximate but fast reconstruction of tracks. For these reasons, Calo b tagging is often used to perform a first event selection which is then refined using PF b tagging. In both cases, tighter track p_T and IP selection criteria compared to the offline algorithms are applied in order to reduce the computation time. This leads to lower track reconstruction efficiency, especially in the region of high track IP, which is one of the main limitations to the online b tagging performance.

In Calo b tagging, a first estimate of the position of the PV along the beam axis is performed by the fast primary vertex (FPV) algorithm [2]. Pixel hits are associated to Calo jets using angular information, and they are projected on the beam line along the jet direction, as illustrated in Fig. 4.10. In this procedure, only the four jets with the highest p_T are considered. The region along the beam axis with the highest concentration of projected hits approximates the position of the PV with a resolution of about 3 mm [2], as shown in Fig. 4.11 (left). Tracks originating close to the estimated PV position and compatible with one of the eight leading-p_T jets in the event are then reconstructed by means of an iterative procedure similar to the one described in Sect. 4.1 using at first pixel hits only (pixel tracks), and subsequently using all hits in the tracker (full tracks). After each step, the position of the PV is refitted using available tracks. This method significantly improves the resolution on the PV position, as shown in Fig. 4.11 (right). The full tracks are then used to reconstruct secondary vertices using the IVF algorithm, as described in Sect. 4.4.

Figure 4.12 compares the performance of the online CSVv2 and DeepCSV taggers, both in the PF and Calo versions, to that of the corresponding offline taggers. The results are obtained using a simulation of $t\bar{t}$ events with 2017 detector conditions. As expected, better b tagging performance is obtained with the PF version of both the CSVv2 and DeepCSV tagger. Furthermore, the online DeepCSV tagger outperforms the online CSVv2 tagger with an increased efficiency of about 10% in absolute terms, compared at the same misidentification probability. However, significantly better performance is achieved by the offline version of the tagger, which can rely on a more efficient track reconstruction. Nonetheless, the transition to the

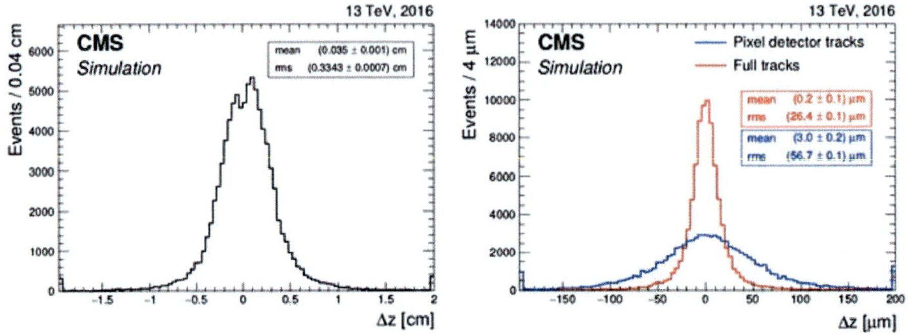

Fig. 4.11 Resolution on the reconstructed position of the primary vertex along the beam axis using the FPV algorithm (left) and after the HLT track fit (right) with pixel and full tracks [2]

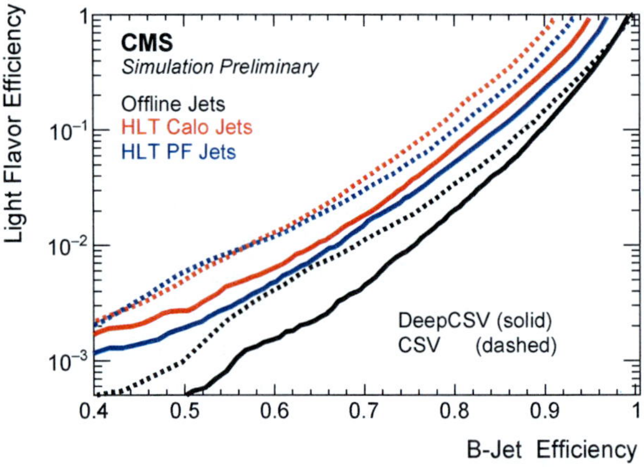

Fig. 4.12 Light jet misidentification probability as a function of the b-tagging efficiency for the online versions of the CSVv2 (dashed lines) and DeepCSV (solid lines) taggers, both with Calo b tagging and PF b tagging, and for the offline taggers [1]

DeepCSV algorithm represents a significant improvement for the b tagging at the HLT, from which several analyses can benefit.

The online b tagging performance is also studied in data by selecting a sample enriched in b jets content, obtained by selecting $t\bar{t}$ event candidates in the final state with one electron and one muon of opposite charge. Each event is required to have an electron of $p_T > 30$ GeV and a muon of $p_T > 20$ GeV, both within the acceptance of the tracker. To further increase the purity of the selected events, the jet with the second highest p_T in the event is required to pass the tight working point of the offline DeepCSV algorithm. The leading jet in the event, which is a b jet with high probability, is then used to study the b tagging performance at the HLT. In the following, this jet is referred to as the probe jet. The studies presented in this thesis are obtained with the 2017 data set, using the 2017 version of the online CSVv2 and the 2018 version of the online DeepCSV.

Figure 4.13 shows the distribution of the online CSVv2 and DeepCSV scores for the probe jet with PF and Calo b tagging. In all cases, good agreement between the data and the simulation is observed. The simulation is split by jet flavour, and the

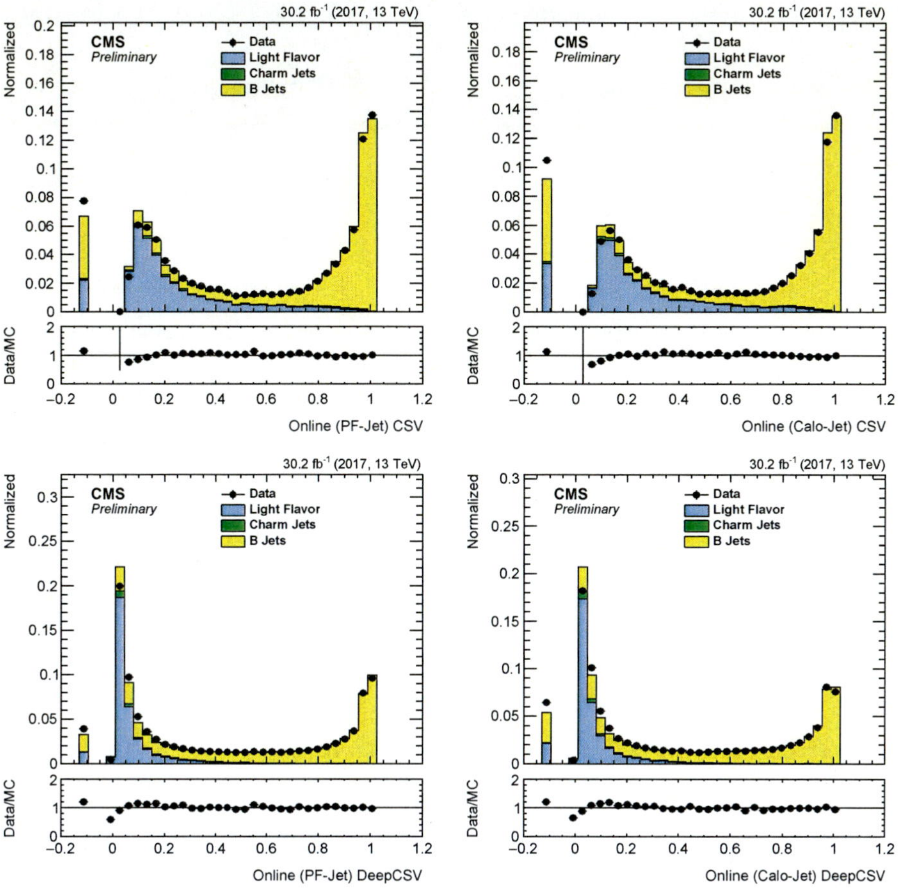

Fig. 4.13 Upper: distribution of the online CSVv2 score in 2017 data and simulation using in PF b tagging (left) and Calo b tagging (right), split by jet flavour. Lower: same distributions for the 2018 version of the online DeepCSV. A sample enriched in b jets is obtained by selecting $t\bar{t}$ events in the $e\mu$ channel [1]

contribution from c jets is found to be negligible. It can be noted that, unlike in the case of the offline taggers, a negative score is assigned to a significantly large fraction of jets. This is a consequence of the lower tracking efficiency at the HLT compared to offline tracking.

The usage of online b tagging in a physics analysis often requires the optimization of the online working points according to the analysis needs. In order to facilitate the process, the efficiency of the online taggers relative to the offline working points are provided to the analysers as a function of the online selection. This quantity is defined as the probability for a jet that passed a given offline working point to pass a certain online selection. The results obtained for the PF versions of CSVv2 and DeepCSV are shown in Fig. 4.14 for the three offline working points. The observed differences

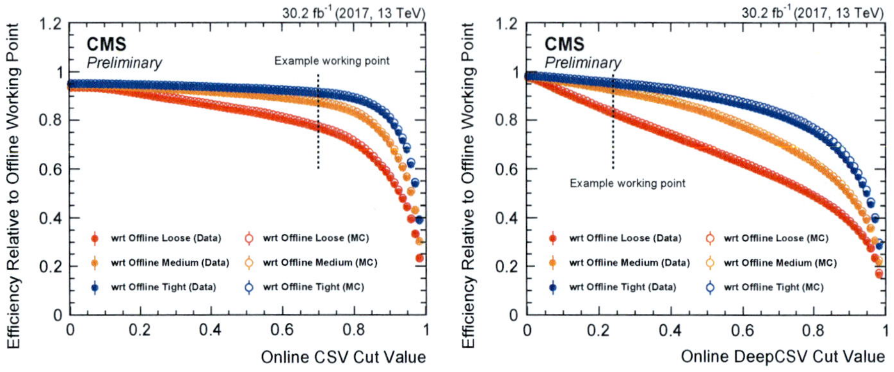

Fig. 4.14 Efficiency of the online PF CSVv2 (left) and DeepCSV (right) relative to the offline operating points as a function of the online selection in 2017 data (full markers) and simulation (hollow markers) [1]

between CSVv2 and DeepCSV have been explained on the basis of the different correlations between the respective online and offline versions of the taggers, which is found to be larger in the case of CSVv2.

4.6 Commissioning of the Double-b Tagger

Searches for physics beyond the standard model often face challenges related to the reconstruction of the decay products of massive bosons produced with large p_T. In fact, when the p_T of the heavy boson is significantly larger than its mass, its decay products merge in a single wide jet and cannot be resolved using standard reconstruction techniques. This is the case, e.g. of searches for heavy resonances decaying into a pair of Higgs bosons. Given the large branching ratio of the $H \to b\bar{b}$ decay channels ($\simeq$58% [20]), these analyses often exploit b tagging techniques to discriminate the signal against background processes [24]. Jets from merged boson decays are typically reconstructed using larger distance parameters, e.g. of 0.8. Additionally, information on the jet substructure can be exploited to discriminate these jets against those initiated by single partons. A typical example of a boosted b tagging algorithm is the double-b tagger [2], which makes use of substructure, lifetime and vertex information to tag the wide jet as a whole.

Since the mass of jets arising from the decay of an electroweak or Higgs boson is usually larger than the mass of a jet initiated by a single parton, the jet mass represents an important observable to discriminate against background processes. However, contributions from pileup, underlying event, and initial-state radiation cause the reconstructed mass of single-parton jets to be significantly larger than the mass of the initial parton. The effect is mitigated using a technique known as jet pruning [25], which is designed to remove the soft and large-angle components from jets, thus preserving the mass of jets originated by heavy bosons while significantly reducing the mass of other jets. The invariant mass of the remaining tracks in the jet is referred to as pruned mass.

Another useful variable, designed to quantify the likelihood of a jet to be originated by N partons, is commonly referred to as N-subjettiness [26]. It is a jet shape observable calculated under the assumption that a jet is made up of N subjets, and is defined as:

$$\tau_N = \frac{1}{d} \sum_k p_{Tk} \min(\Delta R_{1,k}, \Delta R_{2,k}, \ldots, \Delta R_{N,k}) \tag{4.3}$$

where k runs over all the jet constituents, $d = \sum_k p_{Tk}$, R is the jet distance parameter and $\Delta R_{n,k}$ is the distance between the kth constituent and the nth subjet axis. Although τ_N can be calculated with respect to any set of N axes, the subjet axes are defined as those that minimize τ_N, and are referred to as τ-axes. The N-subjettiness is expected to be small if the jet is consistent with the hypothesis of having N or fewer subjets. Therefore, the variables τ_i/τ_j are often used to discriminate between jets originated by i prongs and j prongs.

The double-b tagger is a boosted decision tree (BDT) that combines information from track IP, SVs, and jet substructure to optimally identify wide jets arising from the $H \to b\bar{b}$ decay. SVs are associated to jets within a cone of $\Delta R < 0.7$ around the jet axis, and each SV is matched to the closest τ-axis as shown in Fig. 4.15. The N-subjettiness and the jet mass are not included in the training of the BDT, since they can be used at the analysis level to create control regions to estimate backgrounds. The full list of input variables can be found in Ref. [2]. The double-b tagger can achieve an efficiency of up to 80% for a misidentification probability from light jets of about 10% [2]. Four working points are derived, corresponding to about 75, 65, 45 and 20% signal efficiency for jets with transverse momentum of about 1 TeV. These

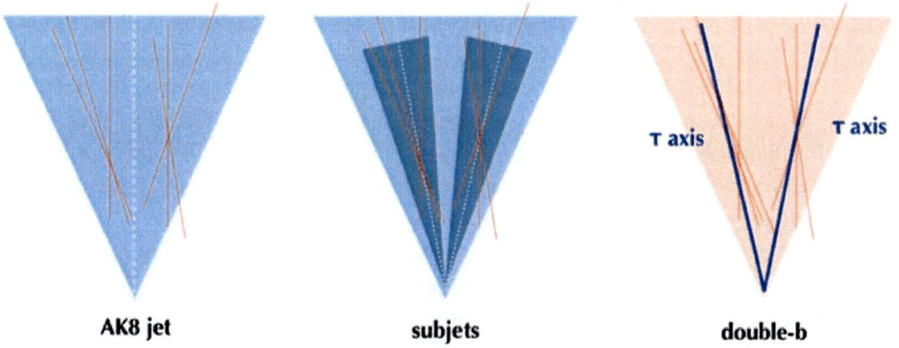

Fig. 4.15 Schematic illustration of the double-b tagger. Secondary vertices are reconstructed independently of jet clustering and matched to subjets. A boosted decision tree is then trained to achieve optimal discrimination against background processes [2]. The wide jet reconstructed with the anti-k_T algorithm with a distance parameter of 0.8 is referred to as AK8 jet

working points are referred to as loose (L), medium 1 (M1), medium 2 (M2), and tight (T), respectively.

One of the main backgrounds to boosted $H \to b\bar{b}$ reconstruction is represented by merged decays of W bosons, especially in the $W \to cs$ and the suppressed $W \to bc$ decay channels. The results presented below represent the first commissioning of the double-b tagger with merged W decays, obtained using proton-proton collisions recorded by CMS in 2016. These results have been obtained in the scope of this thesis, and are documented in Ref. [2]. A pure sample of boosted W boson decays is obtained by selecting $t\bar{t}$ candidate events in the channel where one W boson decays into a muon and a neutrino, while the other decays into a jet of wide area. The event selection requires that the final state contains one isolated muon with $p_T > 50$ GeV and $|\eta| < 2.1$. The selected muon is then used to define two asymmetric hemispheres in the event: the one satisfying the condition $|\phi - \phi_\mu| < 2\pi/3$, where ϕ_μ is the polar angle of the muon, is called leptonic hemisphere, while its complement is referred to as the hadronic hemisphere. Jets reconstructed with the anti-k_T algorithm with a distance parameter of 0.4 (0.8) are referred to as AK4 (AK8) jets. Each hemisphere is required to contain at least one AK4 jet with $p_T > 30$ GeV within the acceptance of the tracker, which represent the b quarks from the decays of the top quarks or additional initial and final state radiation. To ensure high signal purity, at least one AK4 jet in the leptonic hemisphere is required to pass the loose working point of the CSVv2 tagger. In addition, it is required that the hadronic hemisphere contains at least one AK8 jet with $p_T > 250$ GeV, $|\eta| < 2.4$, and pruned mass between 50 and 200 GeV, which represent the merged decay of the W boson. To ensure that the AK8 jet is compatible with a 2-prong decay, it is required that the condition on the subjettiness ratio $\tau_2/\tau_1 < 0.6$ is satisfied. If more than one such jet is found, the one with highest transverse momentum is considered. Backgrounds to this selection consist of single-top and W+jets production. Selected events in simulation are normalized to their predicted cross sections, and the total number of selected events in simulation is scaled to match the one in data. The resulting distributions of the double-b discriminator and of the pruned mass of the AK8 jet are shown in Fig. 4.16. A good level of agreement between the data and the simulation is observed. It can be noted that, unlike for CSVv2 and DeepCSV, the output of the double-b tagger ranges between −1 and 1, and higher scores represent higher probability of a jet to be originated by a $H \to b\bar{b}$ decay. Furthermore, the distribution of the pruned mass of the AK8 jet in Fig. 4.16 (right) exhibits a pronounced peak around the mass of the W boson, indicating that a pure sample of merged W boson decays is selected.

Data-to-MC scale factors (SFs), defined as the ratio between the efficiency of the tagger in data and MC, are derived based on this event selection. For each working point, the efficiency in data is estimated as

$$\epsilon_{\text{data}} = \frac{N^{\text{data}}_{\text{tagged}} - N^{\text{bg,MC}}_{\text{tagged}}}{N^{\text{data}}_{\text{all}} - N^{\text{bg,MC}}_{\text{all}}}, \tag{4.4}$$

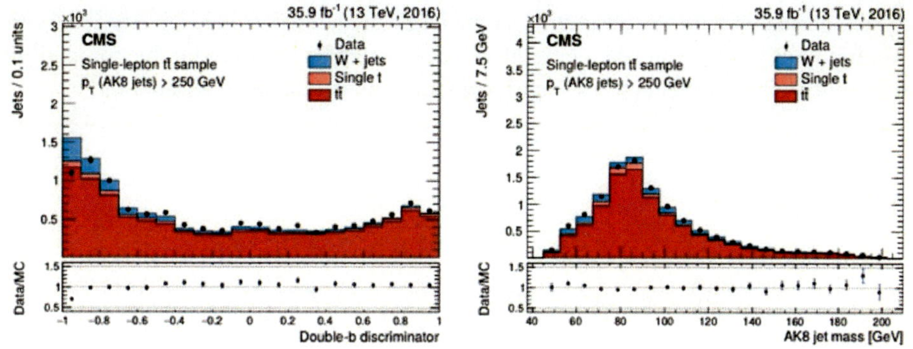

Fig. 4.16 Distributions of the double-b tagger score (left) and of the pruned mass of selected AK8 jets (left) [2]. The simulation is normalized to the data, and the error bars represent the statistical uncertainty in the data

where $N^{\text{data}}_{\text{tagged}}$ is the number of tagged events in the data, $N^{\text{data}}_{\text{all}}$ is the total number of selected events in the data, $N^{\text{bg,MC}}_{\text{tagged}}$ is the number of tagged background events as estimated from the simulation, and $N^{\text{bg,MC}}_{\text{all}}$ is the total number of selected background events as estimated from the simulation. Similarly, the efficiency in the $\text{t}\bar{\text{t}}$ simulation is defined as $\epsilon_{\text{MC}} = N^{\text{t}\bar{\text{t}}}_{\text{tagged}}/N^{\text{t}\bar{\text{t}}}_{\text{all}}$, where $N^{\text{t}\bar{\text{t}}}_{\text{tagged}}$ is the number of tagged $\text{t}\bar{\text{t}}$ events and $N^{\text{t}\bar{\text{t}}}_{\text{all}}$ is the total number of selected $\text{t}\bar{\text{t}}$ events. The scale factor is then defined as $\text{SF}_{\text{top}} = \epsilon_{\text{data}}/\epsilon_{\text{MC}}$. A systematic uncertainty on ϵ_{data} is derived by varying the background normalization by 30%, while the main uncertainty on ϵ_{MC} arises from the modelling of the top quark p_{T} in the MC simulation, and is estimated as explained in Chap. 5. Given the small size of the selected sample, other systematic uncertainties are found to be negligible with respect to the statistical uncertainty of the data. The

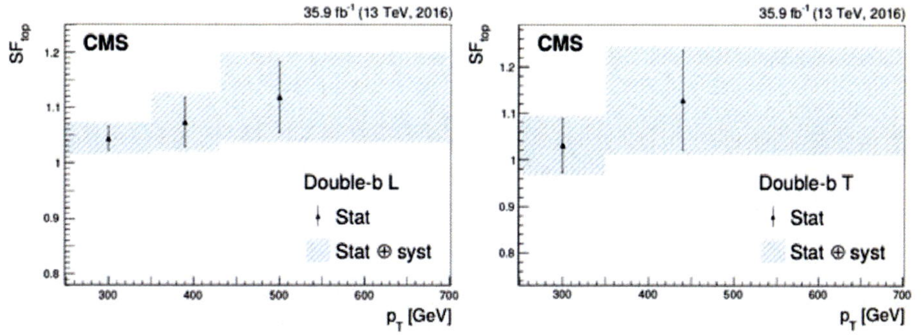

Fig. 4.17 Data-to-simulation scale factors for the double-b tagger misidentification probability from merged W boson hadronic decays in semileptonic $\text{t}\bar{\text{t}}$ events [2]. The vertical error bars correspond to the statistical uncertainty in the data, while the hatched bands represent the combination between the statistical and the systematic uncertainties

measured scale factors as a function of the jet p_T are shown in Fig. 4.17 for the loose and tight working points. The central values of the SFs are found to be compatible with unity, which reflects the good agreement between data and simulation.

References

1. CMS Collaboration (2019) B-jet trigger performance in Run 2, CMS detector performance note CMS-DP-2019-042, CERN, Geneva, November. http://cds.cern.ch/record/2708546
2. CMS Collaboration (2018) Identification of heavy-flavour jets with the CMS detector in pp collisions at 13 TeV. JINST 13:P05011. arXiv:1712.07158
3. CMS Collaboration (2017) Particle-flow reconstruction and global event description with the CMS detector. JINST 12:P10003. arXiv:1706.04965
4. Cauley TMc (2016) Collisions recorded by the CMS detector on 14 Oct 2016 during the high pile-up fill. CMS-PHO-EVENTS-2016-008. https://cds.cern.ch/record/2231915
5. CMS Collaboration (2014) Description and performance of track and primary-vertex reconstruction with the CMS tracker. JINST 9:P10009. arXiv:1405.6569
6. CMS Collaboration (2010) Alignment of the CMS silicon tracker during commissioning with cosmic rays. JINST 5:T03009. arXiv:0910.2505
7. Billoir P, Qian S (1990) Simultaneous pattern recognition and track fitting by the kalman filtering method. Nucl Instrum Meth A 294:219
8. Waltenberger W, Frühwirth R, Vanlaer P (2007) Adaptive vertex fitting. J Phys G 34:343
9. Bartosik N (2016) PF concept in CMS. http://bartosik.pp.ua/hep_sketches/cms_particle_flow
10. Cacciari M, Salam GP, Soyez G (2008) The anti-k_T jet clustering algorithm. JHEP 04:063. arXiv:0802.1189
11. CMS Collaboration (2014) Pileup removal algorithms, Technical report CMS-PAS-JME-14-001, CERN, Geneva
12. CMS Collaboration (2017) Jet energy scale and resolution in the CMS experiment in pp collisions at 8 TeV. JINST 12:P02014. arXiv:1607.03663
13. Cacciari M, Salam GP (2008) Pileup subtraction using jet areas. Phys Lett B 659:119. arXiv:0707.1378
14. Cacciari M, Salam GP, Soyez G (2012) FastJet user manual. Eur Phys J C 72:1896. arXiv:1111.6097
15. CMS Collaboration (2011) Determination of jet energy calibration and transverse momentum resolution in CMS. JINST 6:P11002. arXiv:1107.4277
16. CMS Collaboration (2019) Private CMS page on jet energy corrections. https://twiki.cern.ch/twiki/bin/view/CMS/IntroToJEC. Accessed: August
17. ATLAS Collaboration (2018) Observation of $H \to b\bar{b}$ decays and VH production with the ATLAS detector. Phys Lett B 786:59. arXiv:1808.08238
18. CMS Collaboration (2018) Observation of Higgs boson decay to bottom quarks. Phys Rev Lett 121:121801. arXiv:1808.08242
19. Muller T, Piasecki C, Quast G, Weiser C (2006) Inclusive secondary vertex reconstruction in jets, Technical report CERN-CMS-NOTE-2006-027, CERN, Geneva
20. Particle Data Group, Tanabashi M et al (2018) Review of particle physics. Phys Rev D 98:030001
21. Guest D, Collado J, Baldi P, Hsu S, Urban G, Whiteson D (2016) Jet flavor classification in high-energy physics with deep neural networks. Phys Rev D 94:112002. arXiv:1607.08633
22. CMS Collaboration (2018) Performance of b tagging algorithms in proton-proton collisions at 13 TeV with Phase 1 CMS detector, CMS Detector Performance Note CMS-DP-2018-033, CERN, Geneva. https://cds.cern.ch/record/2627468

23. CMS Collaboration (2018) Performance of the DeepJet b tagging algorithm using 41.9fb^{-1} of data from proton-proton collisions at 13 TeV with Phase 1 CMS detector, CMS Detector Performance Note CMS-DP-2018-058, CERN, Geneva, November http://cds.cern.ch/record/2646773
24. CMS Collaboration (2019) Search for production of Higgs boson pairs in the four b quark final state using large-area jets in proton-proton collisions at $\sqrt{s} = 13\,\text{TeV}$. JHEP 01:040. arXiv:1808.01473
25. Ellis SD, Vermilion CK, Walsh JR (2010) Recombination algorithms and jet substructure: Pruning as a tool for heavy particle searches. Phys Rev D 81:094023. arXiv:0912.0033
26. Thaler J, Van Tilburg K (2011) Identifying boosted objects with N-subjettiness. JHEP 03:015. arXiv:1011.2268

Chapter 5
Simultaneous Measurement of the Top Quark Pair Production Cross Section and the Top Quark Mass

This chapter describes a measurement of the inclusive top quark-antiquark pair ($t\bar{t}$) production cross section at the centre-of-mass energy of 13 TeV, determined simultaneously with the top quark mass in the simulation. The measured cross section is then used in Chap. 6 to extract the strong coupling constant and top quark mass in different renormalization schemes. This chapter is structured as follows: the analysis strategy is described and motivated in Sect. 5.1, where a common procedure used to measure cross sections is also outlined. The criteria used to select the events of interest are discussed in Sect. 5.2, and the treatment of the relevant sources of systematic uncertainties is described in Sect. 5.3. The statistical procedure used to extract the parameters of interest and the extrapolation of the measured cross section to the full phase space are discussed in Sects. 5.4 and 5.5, respectively. The results of the measurement are then presented in Sect. 5.6. Finally, Sect. 5.7 illustrates detailed studies of the impact of the statistical uncertainty in the simulation, performed in the scope of this measurement. This analysis was entirely developed for this thesis following and extending the approach of Ref. [1], and is documented in a paper published in the European Physics Journal [2].

5.1 Measuring the Top Quark Pair Production Cross Section

The production cross section σ of a certain process can be experimentally determined by counting the number of observed events for that process, assuming the integrated luminosity $\mathcal{L}$ of the considered data set is known. In a physics analysis, the geometric acceptance of the detector and the event selection criteria define the visible (fiducial) phase space of the measurement, outside of which events are not detected. Therefore,

M. M. Defranchis, *First Measurement of the Running of the Top Quark Mass*, Springer Theses, https://doi.org/10.1007/978-3-030-90376-3_5

the cross section is only measured in the visible phase space, and is referred to as the visible (fiducial) cross section, σ^{vis}.

Identification criteria are also applied to final-state objects reconstructed in the fiducial phase space in order to reduce the contribution from misreconstructed objects, which further reduces the number of selected events. The probability for an event in the fiducial phase space to pass the full event selection is referred to as selection efficiency, ϵ_{sel}. The fiducial cross section for a given process can be determined experimentally as:

$$\sigma^{\text{vis}} = \frac{N_{\text{data}} - N_{\text{bg}}}{\epsilon_{\text{sel}}\,\mathcal{L}}, \tag{5.1}$$

where N_{data} is the number of selected events in data and N_{bg} is an estimate of the number of selected background events, which can be determined using simulation. The difference between these two quantities estimates the number of events in data of the process of interest. The integrated luminosity is measured as described in Chap. 3, and the selection efficiency can be estimated using simulation. In the case of $\text{t}\bar{\text{t}}$ production, the total cross section is then obtained by extrapolating the visible cross section to the full phase space, using the relation:

$$\sigma = \frac{\sigma^{\text{vis}}}{A_{\text{sel}}\,\text{BR}}, \tag{5.2}$$

where BR is the branching ratio of the considered final state and A_{sel} is the acceptance of the selection, defined as the probability for an event to be produced within the visible phase space. The quantity A_{sel} can only be determined using simulation, and is therefore sensitive to the theoretical assumptions in the MC. This introduces a stronger dependence on the modelling uncertainties (Sect. 5.3), which can be significantly reduced by maximizing the acceptance of the selection.

In physics analyses involving top quarks, the acceptance of the selection also depends on the top quark mass parameter used in the simulation, m_{t}^{MC} [3, 4]. In most analyses, e.g. [5, 6], the cross section is measured at a fixed value of m_{t}^{MC}. In this analysis, instead, the dependence on m_{t}^{MC} is consistently taken into account by performing a simultaneous measurement of the inclusive $\text{t}\bar{\text{t}}$ cross section, $\sigma_{\text{t}\bar{\text{t}}}$, and m_{t}^{MC}, following and extending the method developed in Ref. [1]. In this way, $\sigma_{\text{t}\bar{\text{t}}}$ is determined at the optimal value of m_{t}^{MC} and the correlation between $\sigma_{\text{t}\bar{\text{t}}}$ and m_{t}^{MC} is properly taken into account.

5.2 Data Set, Event Selection, and Monte Carlo Simulation

The measurement is performed using proton-proton collisions data at a centre-of-mass energy of 13 TeV recorded by the CMS detector in 2016, corresponding to an integrated luminosity of 35.9 fb^{-1}. Candidate $\text{t}\bar{\text{t}}$ events are selected in the final state

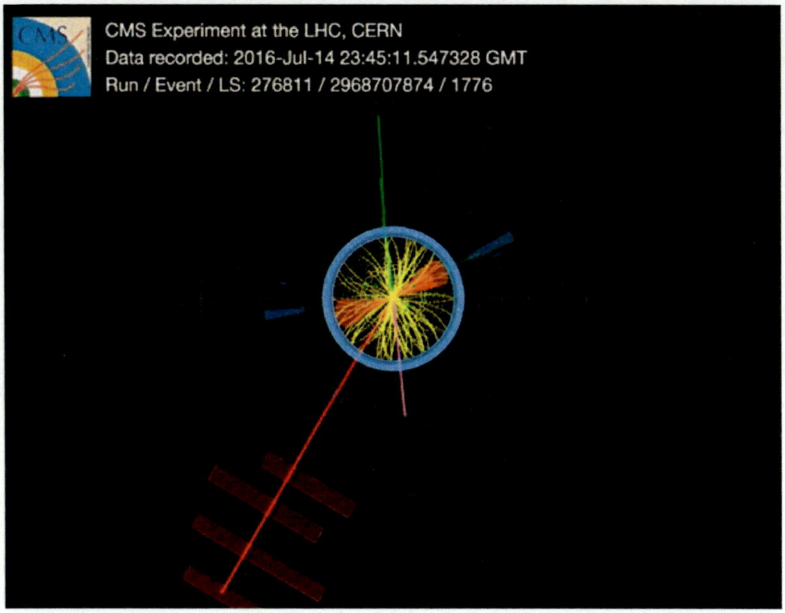

Fig. 5.1 Display of one of the selected events in the $x - y$ plane. The event contains one electron (green) and one muon (red) of opposite charge, two jets with large p_T (orange), and missing transverse energy (purple). The interaction of the reconstructed objects with the CMS detector is also illustrated [7]

where one W boson decays into a muon and a neutrino while the other one decays to an electron and a neutrino. Experimentally, the final state is identified by the presence of an electron and a muon of opposite charge, two b jets, missing transverse energy, and possibly additional jets from QCD radiation. However, in order to maximize the acceptance, events are selected based on requirements on final state leptons only. A display of one of the selected events is shown in Fig. 5.1.

During the data taking, events are collected using a combination of dilepton and single-lepton triggers that require, respectively:

- one electron with $p_T > 12$ GeV and one muon with $p_T > 23$ GeV
- one electron with $p_T > 23$ GeV and one muon with $p_T > 8$ GeV
- one electron with $p_T > 27$ GeV
- one muon with $p_T > 24$ GeV.

All leptons are also required to be within the acceptance of the tracker. The single-lepton triggers are used to recover the inefficiency on the second lepton in dilepton triggers, and increase the overall efficiency from about 90% to almost 100% [8]. All the triggers used in this analysis are not prescaled.

Events selected by the trigger are then reprocessed and further analysed offline. In these events, the presence of an electron and a muon of opposite charge is required, with $|\eta| < 2.4$ and $p_T > 25$ (20) GeV for the leading (subleading) lepton. In events with more than one electron or muon, the two leptons of opposite charge with the highest p_T are considered. If those are both electrons or muons, the event is discarded. Furthermore, since the electron reconstruction performs poorly in the transition region between the barrel and endcap calorimeters, electron candidates are rejected in the range $1.4442 < |\eta| < 1.5660$. Additionally, the invariant mass of the two leptons is required to be larger than 20 GeV in order to suppress the contamination from the QCD multijet background.

Selected leptons are then required to satisfy identification and isolation criteria aimed at reducing the lepton misidentification probability. A relative lepton isolation variable, I_{rel}, is defined as the ratio of the scalar sum of the p_T of the neighbouring particle-flow candidates to the p_T of the lepton. This variable is expected to be small for true, isolated leptons, while it is large for non isolated leptons, e.g. those arising from meson decays in jets. Neighbouring candidates are considered within a radius of $\Delta R < 0.3$ (0.4) of the electron (muon), which is regarded as isolated if $I_{rel} < 0.06$ (0.15). The efficiency for a lepton to pass identification and isolation requirement in simulation is corrected to match the one observed in data by means of data-to-MC scale factors derived using $Z \rightarrow \ell\ell$ events. In addition, all leptons are required to be consistent with originating from the primary vertex in the event.

Jet candidates are reconstructed and calibrated as described in Chap. 4. In this analysis, jets with $p_T > 30$ GeV, $|\eta| < 2.4$, and a distance $\Delta R > 0.4$ from every isolated lepton are considered. However, no requirement on the number of jets is applied. Instead, the jet multiplicity is used to categorize selected events, as discussed in Sect. 5.4. Similarly, b jets are identified, or tagged, using the CSVv2 algorithm [9] described in Chap. 4, and the multiplicity of b tagged jets is used to classify events. A high-purity (tight) working point is chosen for the algorithm, with an average b tagging efficiency in $t\bar{t}$ events of 41%, corresponding to a misidentification probability of 0.1% and 2.2% for light-flavour jets and c jets, respectively [9]. The b tagging efficiency in simulation is corrected to match the one in data using flavour-dependent scale factors measured as a function of the jet p_T. The b tagging scale factors are determined in a muon-enriched QCD sample, while light-jet misidentification corrections are obtained using inclusive QCD events [9]. The b-tagged jet multiplicity in simulation is corrected by using event-by-event weights defined as the ratio of the probability of a given configuration in data (P_{data}) to the probability of the same configuration in MC (P_{MC}). These two quantities can be calculated in terms of the b tagging efficiencies in simulation and the measured SFs as:

$$P_{MC} = \prod_{f=\text{flavour}} \prod_{i=\text{tagged}} \epsilon_i^f(p_T) \prod_{j=\text{untagged}} \left[1 - \epsilon_j^f(p_T)\right], \tag{5.3}$$

$$P_{data} = \prod_{f=\text{flavour}} \prod_{i=\text{tagged}} \text{SF}_i^f(p_T)\, \epsilon_i^f(p_T) \prod_{j=\text{untagged}} \left[1 - \text{SF}_j^f(p_T)\, \epsilon_j^f(p_T)\right], \tag{5.4}$$

where $\epsilon_i^f(p_T)$ and $\text{SF}_i^f(p_T)$ are the b tagging efficiency in MC for jet i and the corresponding scale factor, respectively. These two quantities are determined as a function of the p_T and the flavour f of the jet.

The distributions of the p_T and η of the selected leptons, and the jet and b-tagged jet multiplicity after the event selection are shown in Fig. 5.2 for data and simulated events. Background processes include single-top production in the tW channel, Drell–Yan (DY) events, semi-leptonic $t\bar{t}$ events ($t\bar{t}$ other), W+jets, and diboson production. The tW process is characterized by the same final state as $t\bar{t}$, and is therefore an irreducible background. The DY contribution is represented mainly by leptons originating from $Z \rightarrow \tau^+\tau^-$ decays, while the contamination from semileptonic $t\bar{t}$ and W+jets mainly arises from lepton misidentification. Selected diboson (VV) events can be either WW, WZ, or ZZ events, and are mainly characterized by leptonic decays of the W and Z bosons. The contribution from QCD multi-jet production, instead, was found to be negligible. The $t\bar{t}$, tW, and DY processes are simulated at NLO using the POWHEG program, as described in Chap. 2. In the $t\bar{t}$ and tW simulations, the value of m_t^{MC} is set to 172.5 GeV. The W+jets events are simulated at NLO using MG5_AMC@NLO v2.2.2 [10, 11], while diboson events are generated at leading order using PYTHIA v8.2 [12]. In all simulated samples, the interaction of the final-state particles with the material of the CMS detector is simulated using the GEANT4 software [13–15]. The signal and background simulations are normalized to their predicted cross sections. The distribution of the number of pileup vertices in simulated events is corrected to match the one in the data, which is estimated from the measured instantaneous luminosity assuming a total inelastic pp cross section of 69.2 mb [16]. The bottom-left and bottom-right plots in Fig. 5.2 show that background events are characterized by lower jet and b-tagged jet multiplicities: this fact is used to simultaneously constrain signal and background contributions, as discussed in Sect. 5.4. Overall good agreement between data and simulation is observed within systematic uncertainties. The considered sources of uncertainty are described in the following section.

5.3 Systematic Uncertainties

The hatched bands in Fig. 5.2 represent the effect of systematic uncertainties on the considered final state distributions, which propagates to the final uncertainty in the measured $\sigma_{t\bar{t}}$ and m_t^{MC}. In this analysis, three kinds of systematic uncertainties are considered: experimental uncertainties, which arise from the calibration of the final-state objects used in the analysis; modelling uncertainties, which are related to the assumptions in MC simulation; and background normalization uncertainties, which affect the predicted number of background events. The uncertainty in the centre-of-mass energy was demonstrated to be negligible [17], and is not taken into account in this measurement.

Calibrations of final-state objects are obtained by means of ancillary measurements in which data-to-simulation scale factors (SFs) are determined. These scale factors are used to correct the simulation, and their uncertainty directly propagates to the final result. To assess the impact of these uncertainties, the event selection is

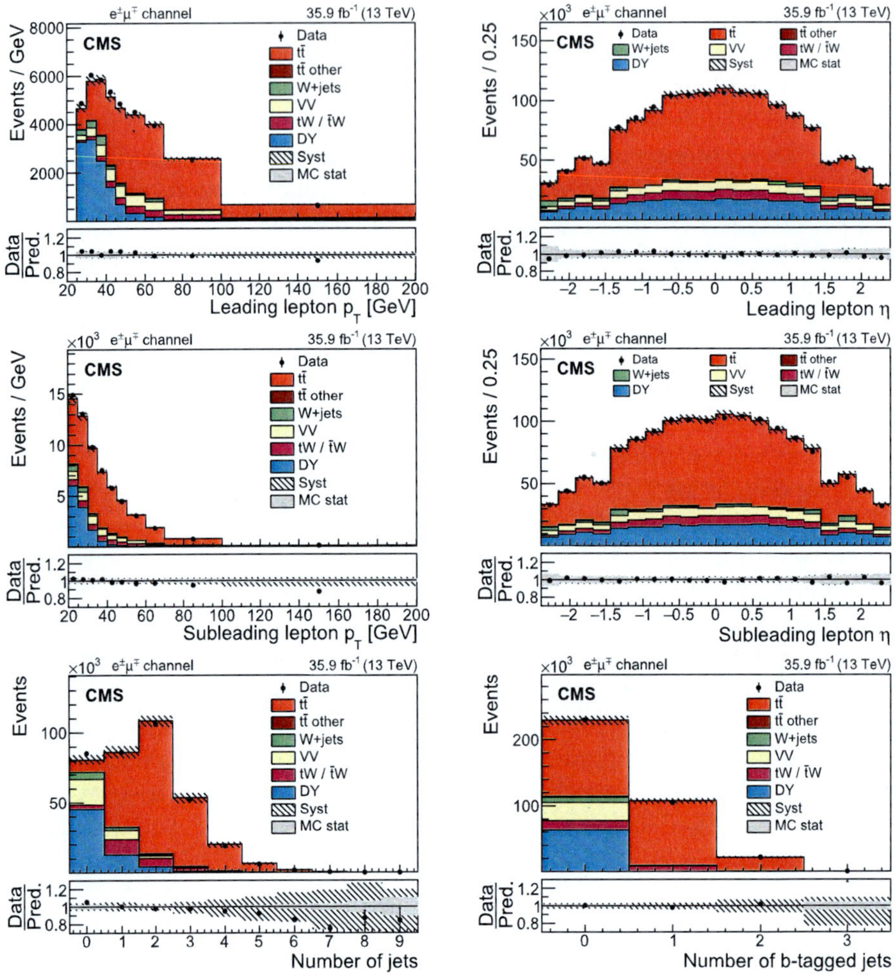

Fig. 5.2 Distributions of the transverse momentum (left) and pseudorapidity (right) of the leading (upper) and subleading (middle) leptons after the event selection for the data (points) and the predictions for the signal and various backgrounds from the simulation (shaded histograms). The lower row shows the jet (left) and b-tagged jet (right) multiplicity distributions. The vertical bars on the points represent the statistical uncertainties in the data. The hatched bands correspond to the systematic uncertainty in the $t\bar{t}$ signal MC simulation. The uncertainties in the integrated luminosity and background contributions are not included. The ratios of the data to the sum of the predicted yields are shown in the lower panel of each figure. Here, the solid grey band represents the contribution of the statistical uncertainty in the MC simulation [2]

repeated after varying the SFs by one standard deviation in both directions. The main sources of experimental uncertainty are described below.

The uncertainty in the jet energy scale (JES) calibration is split into 19 independent sources [18], related, e.g. to the dependence of the calibration on the jet flavour, the MC generator used, the data taking period, the different calibration methods, the modelling of the pileup, and the extrapolation to large jet p_T. The effect of the JES uncertainties is evaluated by varying the p_T of each reconstructed jet within uncertainties, in bins of jet p_T and η. The jet energy resolution (JER) is also corrected by means of η-dependent SFs [19], and the corresponding uncertainty is estimated in the same way. When performing such variations, the reconstructed missing transverse energy is recalculated accordingly. A similar procedure is used for the energy scale and resolution of the leptons.

The main uncertainties affecting the signal and background normalization are related to the integrated luminosity and the lepton identification and isolation scale factors. The uncertainty in the integrated luminosity is measured to be 2.5% [20], while that in the lepton identification is about 1.5% for electrons and 1.2% for muons. A minor source of normalization uncertainty is represented by the corrections to the trigger efficiency, which are determined in bins of lepton p_T and η [8].

Finally, the uncertainties related to the b tagging calibration are estimated by varying the b jet and light jet SFs within their uncertainties, independently. In this analysis, the uncertainty in the b jet SFs is split into 16 different sources, mainly related to the modelling of the soft non-isolated muon used to tag the b jet in the SF measurement [9].

Depending on the source, modelling uncertainties are estimated either by varying the parameters in the simulation, or by comparing the nominal simulation to alternative phenomenological models. To estimate the impact of higher-order effects in the NLO MC, a matrix element (ME) scale uncertainty is obtained by varying the renormalization and factorization scales in the POWHEG simulation of $t\bar{t}$, tW, and DY. The scales are varied independently by a factor of two up and down [21, 22], avoiding combinations for which $\mu_f/\mu_r = 1/4$ or 4. In addition, in order to assess the impact of the choice of the NLO generator, the POWHEG simulation of the $t\bar{t}$ signal is replaced with a sample generated with the MG5_AMC@NLO program with FxFx matching [23].

Similarly, uncertainties related to the parton shower (PS) generator in $t\bar{t}$ and tW events are estimated by varying the initial-state radiation (ISR) and the final-state radiation (FSR) scales by a factor of two [24]. An uncertainty is also assigned to the matching between the ME and the PS, obtained by varying the h_{damp} parameter in POWHEG within its uncertainty, as determined in the MC tuning [25]. Furthermore, an uncertainty related to the UE tuning is estimated by varying the tuning parameters within their uncertainties [25], while an uncertainty related to the modelling of the number of pileup interactions is obtained by changing the inelastic pp cross section by ±4.6% [16].

The assumed b hadron branching fraction (BF) and b quark fragmentation function can affect the reconstructed b jet momentum, and hence the measurement of m_t^{MC}. The uncertainty in the BF is estimated by varying its value by −0.45% and +0.77% [26].

For the fragmentation, the parameters of the Bowler–Lund function [27] determined using LEP data [28, 29] are varied within their uncertainties. An additional uncertainty is obtained by comparing the Bowler–Lund parametrization to the Peterson fragmentation function [30].

Since the uncertainties provided for the NNPDF3.0 PDFs set, which is used to simulate $t\bar{t}$ events, are based on MC replicas and are therefore statistically correlated, PDF uncertainties are estimated using the eigenvectors of the CT14 PDF set [31]. The estimated uncertainty is then scaled from the 95% confidence interval of the provided eigenvectors to a 68% interval.

Previous analyses, e.g. the one described in Ref. [32], indicated that the choice of the colour reconnection (CR) model can have a large effect on direct measurements of the top quark mass. In this analysis, uncertainties on the CR are estimated by comparing between different models. An uncertainty is obtained by enabling early resonance decays (ERD) in PYTHIA, which allows the colour reconnection to happen after the decay of the top quark. Additional uncertainties are obtained by comparing to the "gluon move" model [33], which allows gluons to move between different strings, and the "QCD inspired" model [34], which takes into account effects beyond the leading colour.

Furthermore, an empirical uncertainty is introduced in order to take into account the difference between the top quark p_T spectrum in data and simulation observed in previous analyses [35–37]. The uncertainty is estimated as the difference between the spectrum generated with the POWHEG simulation to the one obtained by reweighting the simulation to the data. The reweighting is performed by means of a smoothly falling exponential function, with parameters determined using results of previous measurements.

The variations relative to the ISR and FSR scales, the matching scale between ME and PS, the UE event tune, the CR model, and the ME generator are estimated by means of dedicated MC samples, while all other variations are obtained by appropriately reweighting the nominal simulation. Furthermore, an additional JES correction is applied to the sample with varied FSR scale in order to account for the different jet response. The same is done for b jets when varying the parameters of the b hadron fragmentation function. In addition, for each systematic variation, the b tagging SFs are corrected by a factor $\epsilon_b^{nom}/\epsilon_b^{var}$, where ϵ_b^{nom} and ϵ_b^{var} are the p_T-dependent b tagging efficiencies in the nominal and varied simulation, respectively. This ensures that the quantity $SF \cdot \epsilon_b$ in Eq. 5.4 approximates the b tagging efficiency in data, irrespective of the systematic variation.

Finally, a 30% normalization uncertainty is conservatively assigned to each background process independently [38]. Since PS scale variations are not available for the utilized DY samples, additional uncertainties are assigned to the jet multiplicity in the DY simulation. In particular, an uncertainty of 5, 10, 30, and 50% for events with exactly 0, 1, 2, and 3 or more jets, respectively, are considered. The first three values are estimated by performing parton shower scale variations in a W+jets simulation at NLO precision, whereas the last one is assigned conservatively [2]. In addition, in order to mitigate the dependence of the result on the b-tagged jet multiplicity in the DY simulation, the normalization uncertainty is applied independently in DY events with different b jet multiplicity.

5.4 The Fit Procedure

The parameters of interest, $\sigma_{\mathrm{t\bar{t}}}$ and $m_{\mathrm{t}}^{\mathrm{MC}}$, are determined by means of a maximum-likelihood fit to multidifferential distributions of final state observables, where systematic uncertainties are treated as nuisance parameters. In the likelihood, the observed number of events in each bin of any observable is assumed to follow a Poisson distribution. The expected number of events in bin i can be expressed as a function of the parameters of the fit as:

$$\nu_i = s_i(\sigma_{\mathrm{t\bar{t}}}, m_{\mathrm{t}}^{\mathrm{MC}}, \vec{\lambda}) + \sum_j b_i^j(\omega_j, \vec{\lambda}), \tag{5.5}$$

where s_i is the expected number of signal events in the bin, and depends on $\sigma_{\mathrm{t\bar{t}}}$, $m_{\mathrm{t}}^{\mathrm{MC}}$, and the nuisance parameters $\vec{\lambda}$, while b_i^j is the number of expected events of background j in that bin, and depends on the nuisance parameters and on the background normalization parameters ω_j. The expected number of tW and semileptonic $\mathrm{t\bar{t}}$ events also depend on $m_{\mathrm{t}}^{\mathrm{MC}}$, although this is not explicitly indicated in Eq. 5.5. The likelihood function is defined as:

$$L(\sigma_{\mathrm{t\bar{t}}}, m_{\mathrm{t}}^{\mathrm{MC}}, \vec{\lambda}, \vec{\omega}) = \prod_i \frac{e^{-\nu_i}\nu_i^{n_i}}{n_i!} \prod_m \pi_m(\lambda_m) \prod_j \pi_j(\omega_j), \tag{5.6}$$

where n_i is the observed number of events in bin i, and $\pi_m(\lambda_m)$ and $\pi_j(\omega_j)$ are priors for the nuisance and normalization parameters, respectively. In this analysis, all nuisance and normalization parameters are assigned a Gaussian prior, with a width of the size of the input uncertainty, while the uncertainty in the integrated luminosity is not included in the likelihood and is estimated separately. The likelihood is maximized with respect to the parameters of interest, the nuisance, and the normalization parameters. This allows the systematic uncertainties to be constrained using data, hence reducing the resulting uncertainty in $\sigma_{\mathrm{t\bar{t}}}$ and $m_{\mathrm{t}}^{\mathrm{MC}}$. The values of the parameters that maximize the likelihood are determined by minimizing the quantity $-2\ln L$ using the MINUIT program [39]. The MINOS algorithm [39] is then used to estimate the uncertainties on the parameters of interest by scanning the value of the likelihood function around its minimum.

In order to constrain the b tagging efficiencies and systematic uncertainties affecting the jet multiplicity (such as the PS scales), events are split into exclusive categories of jet and b-tagged jet multiplicities. Events are classified as having other one, two, or any other number of b-tagged jets. In each of these categories, events are further

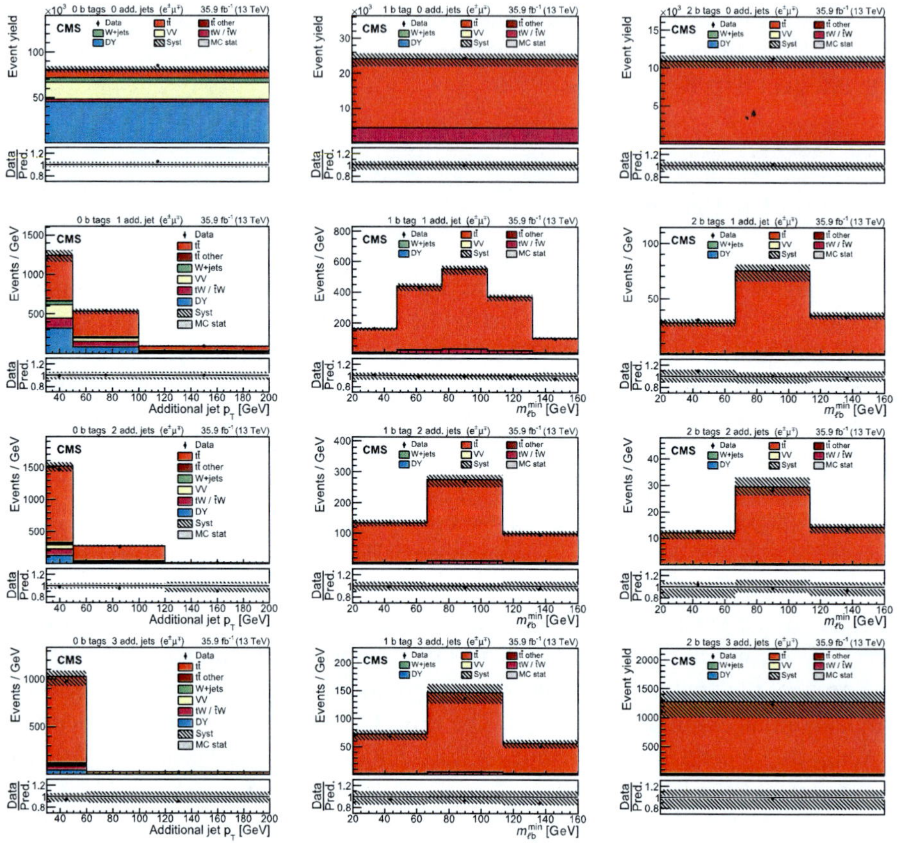

Fig. 5.3 Comparison of data (points) and pre-fit distributions of the expected signal and backgrounds from simulation (shaded histograms) used in the simultaneous fit of $\sigma_{t\bar{t}}$ and m_t^{MC} in the $e^{\mp}\mu^{\pm}$ channel. In the left column events with zero or three or more b-tagged jets are shown. The middle (right) column shows events with exactly one (two) b-tagged jets. Events with zero, one, two, or three or more additional non-b-tagged jets are shown in the first, second, third, and fourth row, respectively. The hatched bands correspond to the total uncertainty in the sum of the predicted yields. The ratios of data to the sum of the predicted yields are shown in the lower panel of each figure. Here, the solid grey band represents the contribution of the statistical uncertainty [2]

classified as having zero, one, two, or more than two non-b-tagged jets. This also allows to create signal- and background-enriched regions that can be used to optimally determine the normalization of the various processes. In each subcategory, a suitable final-state distribution is chosen as input to the fit, as shown in Fig. 5.3. The p_T of the softest selected jet is chosen as observable in events without b-tagged jets and with at least one non-b-tagged jets. This distribution is used to constrain the JES uncertainties at low p_T, where the systematic uncertainties are the largest. In categories without additional non-b-tagged jet and in the category with 2 b-tagged jets and at least 3 additional jets, the event yield is used as input to the fit. This choice is made either to reduce the impact of backgrounds or due to limited statistics in the simulation (Sect. 5.7), depending on the category. For all other categories, the $m_{\ell b}^{min}$ distribution is used as input: this variable is defined as the minimum invariant mass that is found when combining a b-tagged jet with a lepton. This distribution is known to be sensitive to the value of the top quark mass at its end point [1], and is used in this analysis to constrain m_t^{MC}. The effect on the $m_{\ell b}^{min}$ distribution of a ± 3 GeV variation in m_t^{MC} can be found in Appendix A. The binning of the distributions is chosen in order to minimize the effect of the MC statistical uncertainty (Sect. 5.7) while preserving the sensitivity to the parameters of interest. The impact of the main modelling uncertainties on the fit distributions can be found in Appendix A.

In each category of b-tagged jet multiplicity, the effect of the systematic uncertainties on the signal normalization is parametrized by expressing the number of events in the different categories in terms of multinomial probabilities. Based on the fact that two b jets are expected in every $t\bar{t}$ event, the following parametrization is obtained [4]:

$$S_{1b} = \mathcal{L} A_{\ell\ell} \sigma_{t\bar{t}} \epsilon_{\ell\ell} 2\epsilon_b (1 - C_b \epsilon_b), \tag{5.7}$$

$$S_{2b} = \mathcal{L} A_{\ell\ell} \sigma_{t\bar{t}} \epsilon_{\ell\ell} \epsilon_b^2 C_b, \tag{5.8}$$

$$S_{other} = \mathcal{L} A_{\ell\ell} \sigma_{t\bar{t}} \epsilon_{\ell\ell} \left[1 - 2\epsilon_b (1 - C_b \epsilon_b) - \epsilon_b^2 C_b\right]. \tag{5.9}$$

Here S_{1b}, S_{2b}, and S_{other} denote the number of events with one, two, or a different number of b-tagged jets, respectively, $\mathcal{L}$ indicates the integrated luminosity, and $A_{\ell\ell}$ is the acceptance of the selection defined by the kinematic cuts in p_T and η on the two leptons described in Sect. 5.2. Unlike in the case of A_{sel} in Sect. 5.1, the quantity $A_{\ell\ell}$ incorporates the branching ratio of the considered final state, which includes $e^{\mp}\mu^{\pm}$ decays via intermediate τ leptons. Furthermore, $\epsilon_{\ell\ell}$ denotes the efficiency for events in the visible phase space, i.e. within the acceptance of the selection, to pass the full event selection described in Sect. 5.2, ϵ_b is the b-tagging efficiency, and the quantity C_b corrects for any small correlations between the tagging of two b jets in an event. In the fit, the parameters $A_{\ell\ell}$, $\epsilon_{\ell\ell}$, ϵ_b, and C_b are expressed in terms of $\vec{\lambda}$, $\vec{\omega}$, and m_t^{MC}. Therefore, with this parametrization, the dependence of the selection acceptance on m_t^{MC} is fully taken into account. The remaining effects of the systematic uncertainties on the signal simulation are then treated as shape uncertainties, and the quantities s_i in Eq. 5.5 are estimated based on the signal shape and normalization in the corresponding category of b-tagged jet multiplicity. The signal shape is defined as s_i/S_b, where S_b is either S_{1b}, S_{2b}, or S_{other}, depending on the b jet category.

The quantities $A_{\ell\ell}$, $\epsilon_{\ell\ell}$, ϵ_b, C_b, s_i/S_b, and b_i^j and their dependence on the parameters of the fit are estimated using simulation. Templates of the expected final state distributions are obtained for the nominal simulation and for each systematic variation, and are used to estimate the quantities above. A continuous dependence of these parameters on $\vec{\lambda}$ and $\vec{\omega}$ is then derived by interpolating between values obtained with the nominal and varied simulation. Similarly, the dependence on m_t^{MC} in the $t\bar{t}$ and tW simulations is estimated using dedicated samples generated with $m_t^{MC} = 172.5 \pm 3$ GeV. A second-order polynomial is used for the interpolation in the case of two-sided variations, while a linear function is used for one-sided vari-

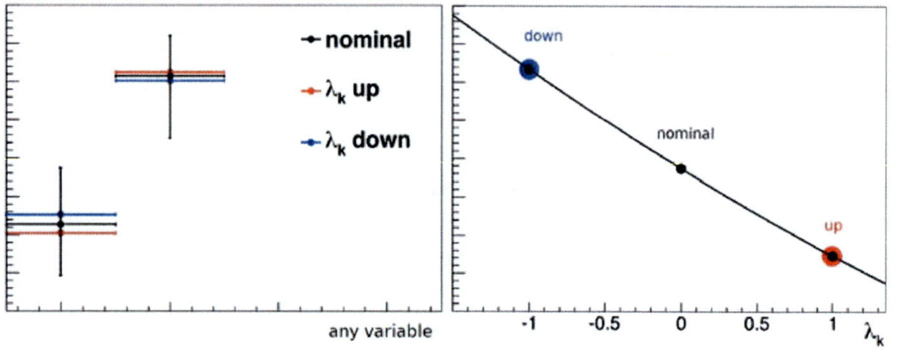

Fig. 5.4 Illustration of the process of template morphing. The impact of a systematic uncertainty on the templates is estimated by varying the corresponding nuisance parameter by one standard deviation (left). A continuous dependence of the considered variable is then obtained by interpolating between the three points by using a second-order polynomial (right) [40]

ations. This process, also called template morphing, is illustrated in Fig. 5.4. Since the parameters $A_{\ell\ell}$, $\epsilon_{\ell\ell}$, ϵ_{b}, and C_{b} are less subject to statistical fluctuations than the bin contents s_i, a more precise parametrization of the signal normalization with respect to $\vec{\lambda}$, $\vec{\omega}$, and $m_{\mathrm{t}}^{\mathrm{MC}}$ is obtained with this method. The values of the efficiencies and correlations for the nominal signal simulation are found to be $\epsilon_{\ell\ell} = 0.49$, $\epsilon_b = 0.30$, and $C_b = 1.00$. The value of $A_{\ell\ell}$ is found to be 1.61%, which corresponds to $A_{\mathrm{sel}} \simeq 80\%$.

The post-fit distributions of the observables used as input to the fit are shown in Fig. 5.5. Improved agreement between data and simulation is observed, and the impact of systematic uncertainties is significantly reduced. The impact of the main modelling uncertainties on the input distributions after the fit to the data are shown in Appendix A, and a closure test of the fit procedure can be found in Appendix B.

5.5 Extrapolation to the Full Phase Space

The total cross section is estimated by correcting the measured visible cross section for the acceptance of the selection as discussed in Sect. 5.1. This procedure is commonly referred to as extrapolation. In this analysis, the dependence of the acceptance on the nuisance parameters incorporated in the fit, where the parameters $\vec{\lambda}$ are constrained within the visible phase space. However, the validity of these constraints outside the visible phase space cannot be assumed, and additional uncertainties must be assigned to the extrapolation. The procedure used to estimate the extrapolation uncertainties is described in the following.

For each relevant source of modelling uncertainty, the acceptance as determined in the fit is re-evaluated by varying the corresponding nuisance parameter by one pre-fit standard deviation in both directions with respect to its pre-fit value, while keeping

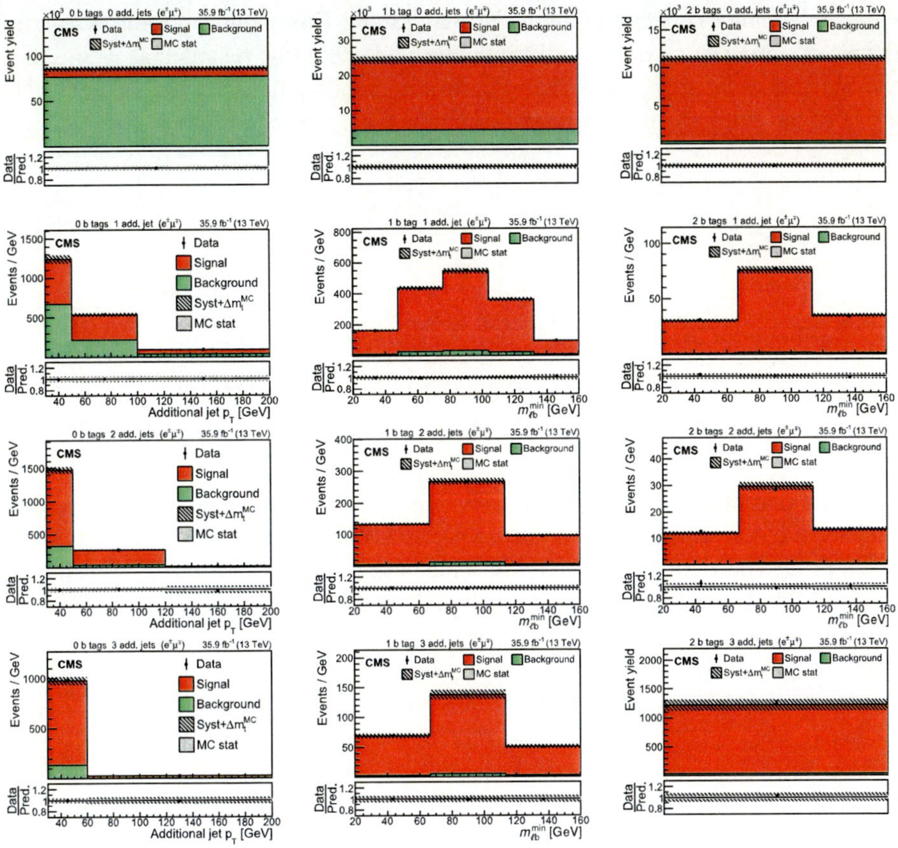

Fig. 5.5 Comparison of data (points) and post-fit distributions of the expected signal and backgrounds from simulation (shaded histograms) used in the simultaneous fit of $\sigma_{t\bar{t}}$ and m_t^{MC} in the $e^{\mp}\mu^{\pm}$ channel. In the left column events with zero or three or more b-tagged jets are shown. The middle (right) column shows events with exactly one (two) b-tagged jets. Events with zero, one, two, or three or more additional non-b-tagged jets are shown in the first, second, third, and fourth row, respectively. The hatched bands correspond to the total uncertainty in the sum of the predicted yields and include the contribution from the top quark mass (Δm_t^{MC}). The ratios of data to the sum of the predicted yields are shown in the lower panel of each figure. Here, the solid grey band represents the contribution of the statistical uncertainty [2]

all other nuisance parameters fixed to their post-fit values. The up/down extrapolation uncertainty corresponding to source i, indicated with $\Delta_i^{\pm}$, is then estimated as:

$$\frac{\Delta_i^{\pm}}{\sigma_{t\bar{t}}} = \left| 1 - \frac{A^{fit}}{A_i^{\pm}} \right| , \tag{5.10}$$

where A^{fit} is the acceptance evaluated at the best-fit value of all the nuisance parameters, and $A_i^{\pm}$ is the acceptance corresponding to the up/down variation of the nuisance parameter corresponding to source i. The uncertainties relative to the different modelling uncertainties are then combined in quadrature. This method introduces a partial double-counting of modelling uncertainties, and should be regarded as a conservative but consistent approach to incorporate extrapolation uncertainties in the fit. The effect of the extrapolation uncertainties on the measured $\sigma_{t\bar{t}}$ is shown in Table 5.1.

5.6 Results

The values of $\sigma_{t\bar{t}}$ and m_t^{MC} as measured to be:

$$\sigma_{t\bar{t}} = 815 \pm 2\,(\text{stat}) \pm 29\,(\text{syst}) \pm 20\,(\text{lum})\,\text{pb},$$
$$m_t^{MC} = 172.33 \pm 0.14\,(\text{stat}) \pm^{0.66}_{0.72}\,(\text{syst})\,\text{GeV},$$

and their correlation is estimated to be 12%. Here, the uncertainties indicate the statistical, systematic, and luminosity uncertainty. The systematic uncertainty also includes the contribution from the MC statistics, which is estimated as described in Sect. 5.7. The result for $\sigma_{t\bar{t}}$ represents one of the most precise measurements obtained with CMS data at $\sqrt{s} = 13$ TeV. The result for m_t^{MC} is the most precise direct top quark mass measurement performed in the dileptonic channel, to date, and has a similar precision to those obtained in the semileptonic [32] and fully-hadronic [41] channels. The measured value of $\sigma_{t\bar{t}}$ corresponds to a visible cross section of:

$$\sigma_{t\bar{t}}^{vis} = 12.86 \pm 0.03\,(\text{stat}) \pm 0.43\,(\text{syst}) \pm 0.32\,(\text{lum})\,\text{pb}.$$

The normalized pulls and constraints on the nuisance parameters relative to the modelling uncertainties are shown in Fig. 5.6, while those on all other nuisance parameters are shown in Fig. 5.7. The constraint on a nuisance parameter is the ratio between its post-fit and pre-fit uncertainty, while the normalized pull is defined as the difference between its post-fit and pre-fit value in units of the pre-fit uncertainty. The vast majority of the pulls is found to be smaller than one standard deviation, and strong constraints are only observed for the nuisance parameters related to uncertainties estimated using independent MC samples. This effect is partly artificial, and is accounted for by the MC statistical uncertainty, as explained in Sect. 5.7.

The contributions of the different sources of systematic uncertainties to the total uncertainty in $\sigma_{t\bar{t}}$ (and $\sigma_{t\bar{t}}^{vis}$) and m_t^{MC} are shown in Tables 5.1 and 5.2, respectively. The impact of similar sources (e.g. the different sources of JES and b tagging uncertainties, or the various PDF eigenvectors) are evaluated simultaneously by repeating the fit while keeping the corresponding nuisance parameters fixed to their post-fit values. The impact of the considered sources is approximated by the difference in quadrature between the full uncertainty and the one obtained with this procedure. In this estimate, the correlations between the considered subset of parameters are taken into account. However, the sum in quadrature of the estimated contributions does not correspond to the total uncertainty, as the different estimates are not statistically

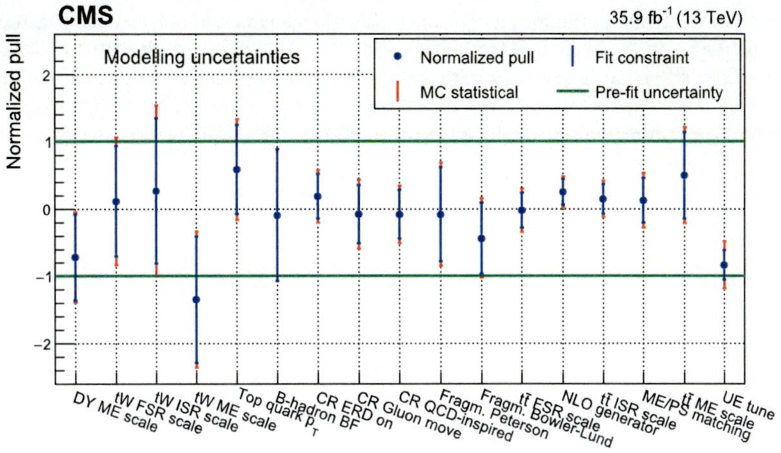

Fig. 5.6 Normalized pulls and constraints of the nuisance parameters related to the modelling uncertainties for the simultaneous fit of $\sigma_{t\bar{t}}$ and m_t^{MC}. The markers denote the fitted value, while the inner vertical bars represent the constraint and the outer vertical bars denote the additional uncertainty as determined from pseudo-experiments. The constraint is defined as the ratio of the post-fit uncertainty to the pre-fit uncertainty of a given nuisance parameter, while the normalized pull is the difference between the post-fit and the pre-fit value of the nuisance parameter normalized to its pre-fit uncertainty. The horizontal lines at ± 1 represent the pre-fit uncertainty [2]

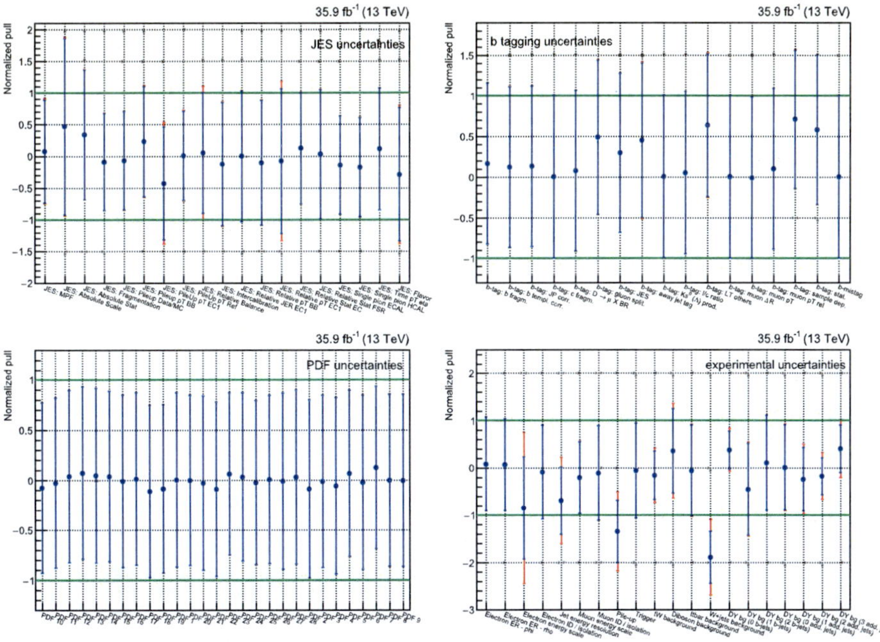

Fig. 5.7 The same as Fig. 5.6 but for JES (upper left), b tagging (upper right), PDF (lower left), and other experimental uncertainties (lower right)

Table 5.1 The relative uncertainties in $\sigma_{t\bar{t}}^{vis}$ and $\sigma_{t\bar{t}}$ and their sources, as obtained from the template fit. The uncertainty in the integrated luminosity and the MC statistical uncertainty are determined separately. The individual uncertainties are given without their correlations, which are however accounted for in the total uncertainty. Extrapolation uncertainties only affect $\sigma_{t\bar{t}}$. For extrapolation uncertainties, the $\pm$ notation is used if a positive variation produces an increase in $\sigma_{t\bar{t}}$, while the $\mp$ notation is used otherwise [2]

Source	Uncertainty [%]
Trigger	0.4
Lepton ident./isolation	2.2
Muon momentum scale	0.2
Electron momentum scale	0.2
Jet energy scale	0.7
Jet energy resolution	0.5
b tagging	0.3
Pileup	0.3
$t\bar{t}$ ME scale	0.5
tW ME scale	0.7
DY ME scale	0.2
NLO generator	1.2
PDF	1.1
m_t^{MC}	0.4
Top quark p_T	0.5
ME/PS matching	0.2
UE tune	0.3
$t\bar{t}$ ISR scale	0.4
tW ISR scale	0.4
$t\bar{t}$ FSR scale	1.1
tW FSR scale	0.2
b quark fragmentation	1.0
b hadron BF	0.2
Colour reconnection	0.4
DY background	0.8
tW background	1.1
Diboson background	0.3
W+jets background	0.3
$t\bar{t}$ background	0.2
Statistical	0.2
Integrated luminosity	2.5

(continued)

Table 5.1 (continued)

Source	Uncertainty [%]
MC statistical	1.2
Total $\sigma_{t\bar{t}}^{vis}$ uncertainty	4.2
Extrapolation uncertainties	
$t\bar{t}$ ME scale	$\mp^{0.4}_{<0.1}$
PDF	$\pm^{0.8}_{0.6}$
Top quark p_T	$\pm^{0.2}_{0.3}$
$t\bar{t}$ ISR scale	$\mp^{0.2}_{<0.1}$
$t\bar{t}$ FSR scale	± 0.1
UE tune	<0.1
m_t^{MC}	$\mp^{0.2}_{0.3}$
Total $\sigma_{t\bar{t}}$ uncertainty	$^{+4.3}_{-4.2}$

independent. The main sources of uncertainties in the measured $\sigma_{t\bar{t}}$ are related to the lepton identification efficiency and the integrated luminosity, which directly affect the normalization of the various processes. In the case of m_t^{MC}, the main sources of uncertainties are related to the JES and the MC statistics, which mainly affect the shape of the considered distributions. The impact of each individual source of systematic uncertainty is summarized in Figs. 5.8 and 5.9 for $\sigma_{t\bar{t}}$ and m_t^{MC}, respectively. The contributions are estimated by varying the corresponding nuisance parameters by one post-fit standard deviation, taking the correlations into account. Also in this case, the estimates are not statistically independent and their combination does not correspond to the total uncertainty.

The measurement of m_t^{MC} is cross-checked by performing an alternative fit without event categorization, where a single $m_{\ell b}^{min}$ distribution is used as input. In this case, events are required to contain at least one b-tagged jet. The $m_{\ell b}^{min}$ distribution before and after the fit to the data is shown in Fig. 5.10, and the measured value of the top mass is:

$$m_t^{MC} = 171.92 \pm 0.13\,(\text{stat})\,^{+0.76}_{-0.77}\,(\text{syst})\,\text{GeV}.$$

This result is compatible with the one of the nominal fit within the uncorrelated part of the uncertainty, which is estimated to be at least 0.54 GeV. In this case, the nuisance parameters are only mildly pulled or constrained, as shown in Fig. 5.11, e.g. for the modelling uncertainties. Furthermore, the impact of the MC statistical uncertainty is smaller than in the nominal fit, and amounts to 0.14 GeV. This is expected, since a single differential distribution is used. As in the nominal fit, the dominant systematic uncertainty is related to the JES, which amounts to 0.67 GeV, while other relevant sources of uncertainties include the JER (0.16 GeV), the ME/PS matching (0.19 GeV), the modelling of the top quark p_T (0.17 GeV), and the statistical

Table 5.2 The absolute uncertainties in m_t^{MC} and their sources, from the simultaneous fit of $\sigma_{t\bar{t}}$ and m_t^{MC}. The MC statistical uncertainty is determined separately. The individual uncertainties are given without their correlations, which are however accounted for in the total uncertainty [2]

Source	Uncertainty [GeV]
Trigger	0.02
Lepton ident./isolation	0.02
Muon momentum scale	0.03
Electron momentum scale	0.10
Jet energy scale	0.57
Jet energy resolution	0.09
b tagging	0.12
Pileup	0.09
$t\bar{t}$ ME scale	0.18
tW ME scale	0.02
DY ME scale	0.06
NLO generator	0.14
PDF	0.05
$\sigma_{t\bar{t}}$	0.09
Top quark p_T	0.04
ME/PS matching	0.16
UE tune	0.03
$t\bar{t}$ ISR scale	0.16
tW ISR scale	0.02
$t\bar{t}$ FSR scale	0.07
tW FSR scale	0.02
b quark fragmentation	0.11
b hadron BF	0.07
Colour reconnection	0.17
DY background	0.24
tW background	0.13
Diboson background	0.02
W+jets background	0.04
$t\bar{t}$ background	0.02
Statistical	0.14
MC statistical	0.36
Total m_t^{MC} uncertainty	$^{+0.68}_{-0.73}$

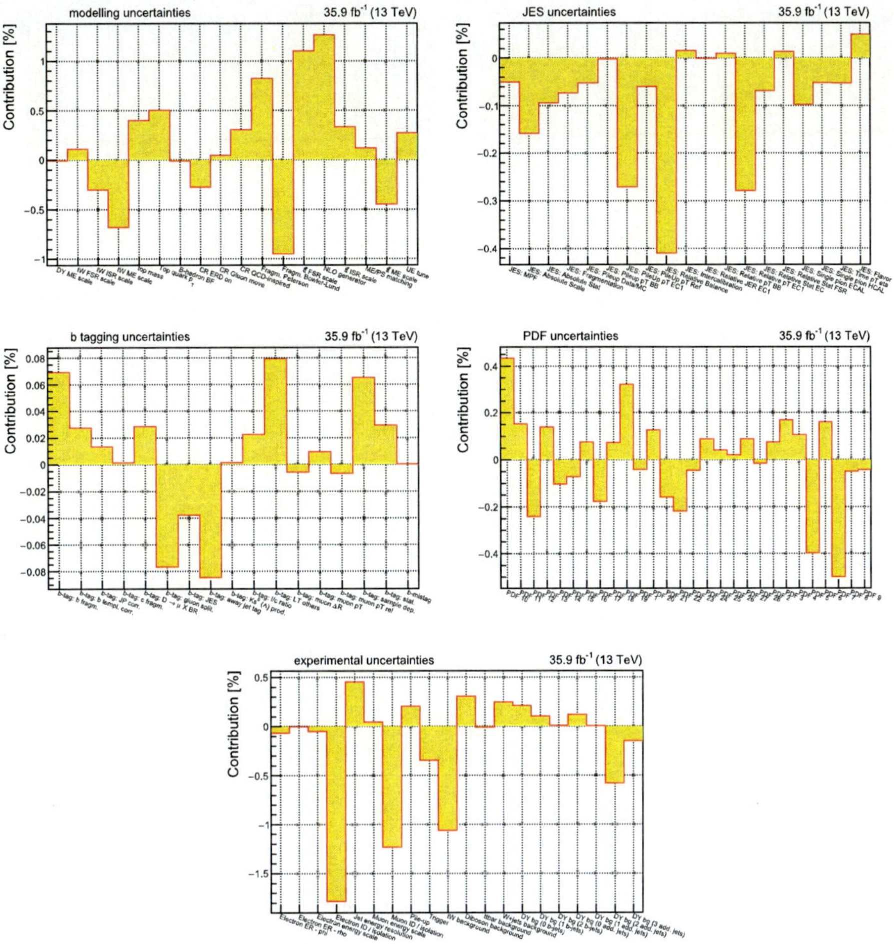

Fig. 5.8 Contributions of the individual sources of systematic uncertainty to the total uncertainty in $\sigma_{t\bar{t}}$. The plots refer to modelling (upper left), JES (upper right), b tagging (middle left), PDF (middle right), and other experimental uncertainties (lower). A positive sign is assigned to the contribution if an increase in the corresponding nuisance parameter determined an increase of $\sigma_{t\bar{t}}$, while a negative sign is assigned otherwise

uncertainty of the data (0.13 GeV). The impact of the statistical uncertainty is larger in this case, since less events are used in the fit due to the requirement on the number of b-tagged jets. Further details on the result of the cross-check fit can be found in Appendix C.

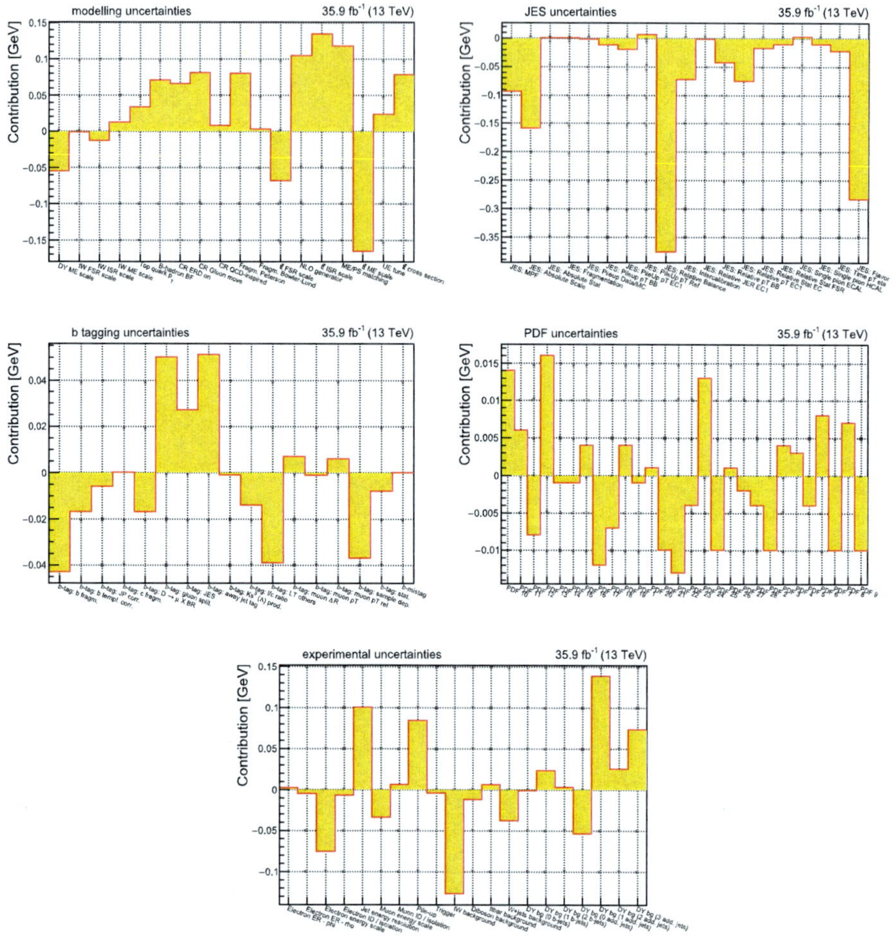

Fig. 5.9 The same as Fig. 5.8, but for the uncertainty in m_t^{MC}

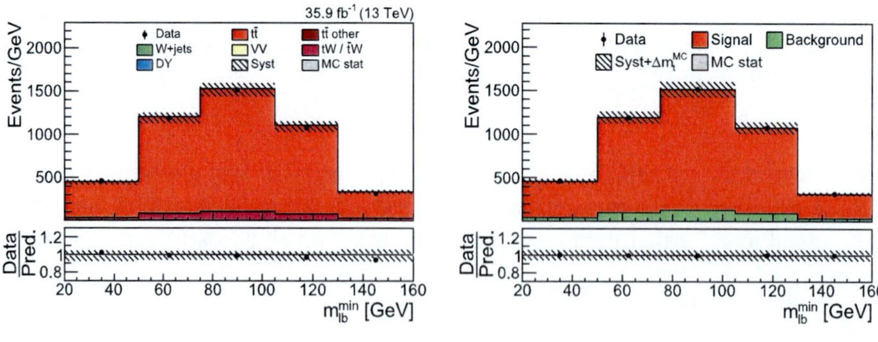

Fig. 5.10 The $m_{\ell b}^{min}$ distribution used in the cross-check fit of m_t^{MC} before (left) and after (right) the fit. In this case, the distribution contains all events with at least one b tagged jet, without categorization

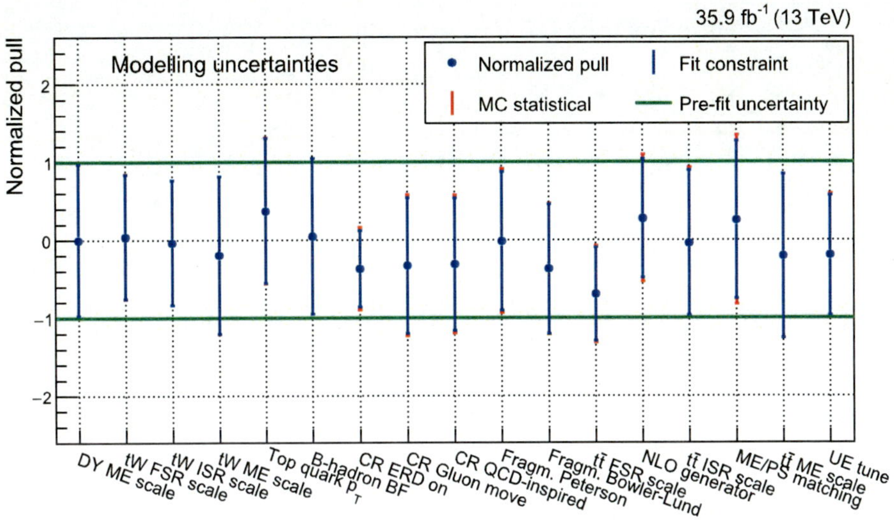

Fig. 5.11 The same as Fig. 5.6, but for the cross-check fit of m_t^{MC} obtained using a single $m_{\ell b}^{min}$ distribution without event categorization

5.7 Impact of Statistical Uncertainties in the MC Simulation

Fits making use of MC templates, as the one described in this chapter, rely on the shape and normalization obtained using simulations in order to constrain the nuisance parameters and to determine the parameters of interest. Although in most cases, including this analysis, the number of events in the simulation is sufficiently larger than the one in data, the statistical uncertainty in the templates can alter the shape of the predicted distributions and introduce significant uncertainties in the template morphing. This can lead to unphysical constraints of the nuisance parameters, resulting in an underestimation of systematic uncertainties. The impact of the MC statistical uncertainty is often estimated by means of bin-by-bin (BBB) fluctuations, which implies fluctuating the content of each bin according to Poisson statistics. However, this procedure does not take into account the uncertainty in the template morphing, as illustrated in Fig. 5.12 (left). A method to estimate this uncertainty has been developed for the purpose of this analysis, and is presented in the following.

Depending on the uncertainty source, the templates corresponding to a systematic variation (also called varied templates) are obtained in three different ways:

1. by replacing the nominal MC simulation with an alternative one, as in the case of m_t^{MC}, CR models, PS scales, ME/PS matching, choice of NLO generator, and UE event tune;
2. by varying the kinematic properties of reconstructed final state objects, as in the case of JES and JER uncertainties and of the energy scale and resolution of

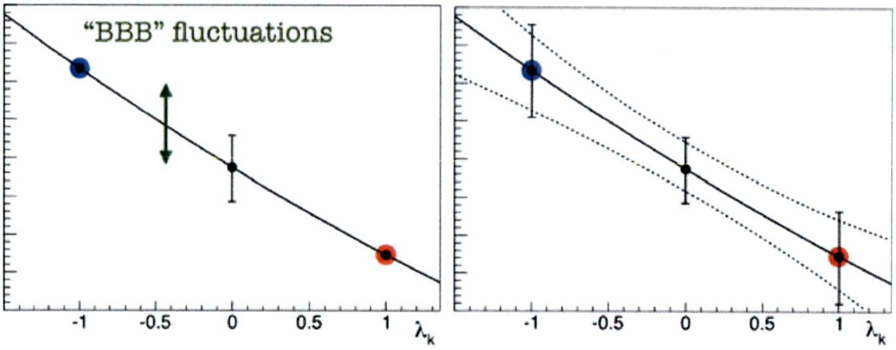

Fig. 5.12 Bin-by-bin fluctuations (left) are often used to estimate the impact on the final result of the statistical uncertainty in the templates, although they do not take into account the uncertainty in the template morphing (right) [40]

electron and muons, which are estimated by varying the reconstructed p_T of the object by the corresponding uncertainty;

3. by reweighting events, as in all other cases.

In the last case, the nominal and varied templates contain the same events, although with different weights. Provided that the weights do not differ significantly, as in the case of this analysis, the statistical uncertainties in corresponding bins can be considered as fully correlated and the template morphing is almost exact. In the first case, instead, the templates contain events generated independently. Therefore, the statistical uncertainties in corresponding bins are fully uncorrelated and the template morphing is only approximate, as illustrated in Fig. 5.12 (right). In the second case, instead, variations of the kinematic properties of final state objects can introduce event migrations across the different bins, and the statistical uncertainty in the nominal and varied templates are only partially correlated. In order to minimize the migrations, and hence maximize the correlations, a coarse binning is chosen for the distributions used in this analysis, as shown in Fig. 5.3. For this reason, the nominal and varied templates are assumed to be fully correlated in the case of JES, JER, and lepton energy scale variations.

In this analysis, the impact of the uncertainty in the template morphing is estimated by means of a toy-experiment procedure. Toy templates are generated by fluctuating the bin content of each distribution according to Poisson statistics, based on the number of simulated events in that bin. The templates obtained by event reweighting or by varying the kinematic properties of the final state objects are fluctuated consistently with the nominal template, while those derived from the alternative samples are fluctuated independently. In this way, the correlation between the nominal template and the varied ones is taken into account. At each iteration, the template dependencies are re-derived, and the fit to the data is repeated. Furthermore, in order to take into account the impact on the correlations between the systematic uncertainties, all templates are fluctuated simultaneously.

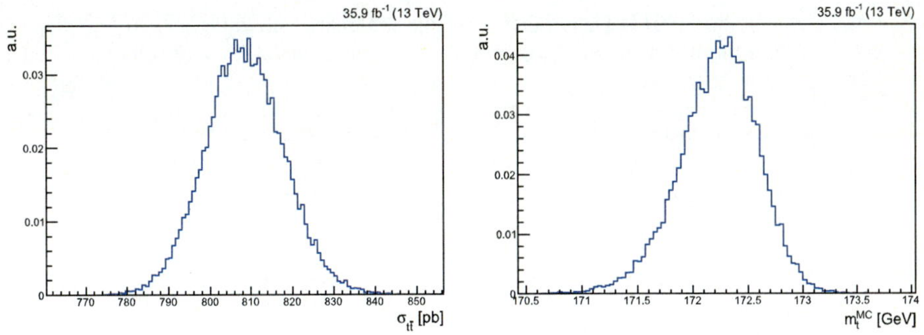

Fig. 5.13 Best-fit values of $\sigma_{t\bar{t}}$ (left) and m_t^{MC} (right) obtained using toy templates. The spread in the obtained distributions represents the impact of the statistical uncertainty in the simulation

At each iteration, the best-fit values of all the parameters in the fit are recorded. The distributions of the values of $\sigma_{t\bar{t}}$ and m_t^{MC} obtained after 30 thousand iterations are shown in Fig. 5.13, and exhibit a quasi-Gaussian shape centred at the measured values of $\sigma_{t\bar{t}}$ and m_t^{MC}. The root mean square (RMS) of the distribution estimates the uncertainty due to the limited MC statistics, and is added in quadrature to the total uncertainty obtained in the fit. In this way, the effect of any possible unphysical constraint is consistently estimated and corrected for. As shown in Tables 5.1 and 5.2, the effect is found to be significant in the case of m_t^{MC}, which is determined from the shape of final state distributions, while it is less pronounced for $\sigma_{t\bar{t}}$, which mainly depends on the overall normalization.

Similarly, the RMS of the best-fit value of each nuisance parameter is added in quadrature to its post-fit uncertainty, and is represented by the outer vertical bands in Figs. 5.6 and 5.7. A non-negligible impact is also observed for the nuisance parameters related to uncertainties determined by event reweighting, although the effect is less significant compared to the other cases. This is due to the fact that the statistical uncertainty in the templates also affects the measured correlations, and the effect propagates to all the parameters of the fit. The distributions of the pulls and constraints on the main modelling uncertainties obtained using toy templates can be found in Appendix D.

References

1. Kieseler J (2015) Measurement of top-quark pair production cross sections and calibration of the top-quark Monte-Carlo mass using LHC Run I proton-proton collision data at $\sqrt{s} = 7$ and 8 TeV with the CMS experiment, PhD thesis. http://cds.cern.ch/record/2133096. https://doi.org/10.1007/978-3-319-40005-1
2. CMS Collaboration (2019) Measurement of the $t\bar{t}$ production cross section, the top quark mass, and the strong coupling constant using dilepton events in pp collisions at $\sqrt{s} = 13$ TeV. Eur Phys J C 79:368. arXiv:1812.10505

3. CMS Collaboration (2014) Determination of the top-quark pole mass and strong coupling constant from the $t\bar{t}$ production cross section in pp collisions at $\sqrt{s} = 7\,\text{TeV}$. Phys Lett B 728:496. arXiv:1307.1907
4. CMS Collaboration (2016) Measurement of the $t\bar{t}$ production cross section in the $e\mu$ channel in proton-proton collisions at $\sqrt{s} = 7$ and 8 TeV. JHEP 08:029. arXiv:1603.02303
5. ATLAS Collaboration (2019) Measurement of the $t\bar{t}$ production cross-section and lepton differential distributions in $e\mu$ dilepton events from pp collisions at $\sqrt{s} = 13\,\text{TeV}$ with the ATLAS detector, Technical report ATLAS-CONF-2019-041, CERN, Geneva. https://cds.cern.ch/record/2686255
6. CMS Collaboration (2017) Measurement of the $t\bar{t}$ production cross section using events with one lepton and at least one jet in pp collisions at $\sqrt{s} = 13\,\text{TeV}$. JHEP 09:051. arXiv:1701.06228
7. CMS Collaboration (2019) Watching the top quark mass run. https://cms.cern/news/watching-top-quark-mass-run, Accessed: December 2019
8. Arndt TM (2019) Precision measurement of the top quark pair production cross section at $\sqrt{s} = 13$TeV with the CMS detector, PhD thesis. http://cds.cern.ch/record/2693956
9. CMS Collaboration (2018) Identification of heavy-flavour jets with the CMS detector in pp collisions at 13 TeV. JINST 13:P05011. arXiv:1712.07158
10. Alwall J, Frederix R, Frixione S, Hirschi V, Maltoni F, Mattelaer O et al (2014) The automated computation of tree-level and next-to-leading order differential cross sections, and their matching to parton shower simulations. JHEP 07:079. arXiv:1405.0301
11. Frixione S, Webber BR (2002) Matching NLO QCD computations and parton shower simulations. JHEP 06:029. arXiv:hep-ph/0204244
12. Sjöstrand T, Ask S, Christiansen JR, Corke R, Desai N, Ilten P et al (2015) An introduction to PYTHIA 8.2. Comput Phys Commun 191:159. arXiv:1410.3012
13. GEANT4 Collaboration (2003) GEANT4: a simulation toolkit. Nucl Instrum Meth A 506:250
14. GEANT4 Collaboration (2006) GEANT4 developments and applications. IEEE Trans Nucl Sci 53:270
15. GEANT4 Collaboration (2016) Recent developments in GEANT4. Nucl Instrum Meth A 835:186
16. ATLAS Collaboration (2016) Measurement of the inelastic proton-proton cross section at $\sqrt{s} =$ 13 TeV with the ATLAS detector at the LHC. Phys Rev Lett 117:182002. arXiv:1606.02625
17. Todesco E, Wenninger J (2017) Large Hadron Collider momentum calibration and accuracy. Phys Rev Accel Beams 20:081003
18. CMS Collaboration (2017) Jet energy scale and resolution in the CMS experiment in pp collisions at 8 TeV. JINST 12:P02014. arXiv:1607.03663
19. CMS Collaboration (2017) Jet algorithms performance in 13 TeV data, CMS Physics Analysis Summary CMS-PAS-JME-16-003, CERN, Geneva. http://cds.cern.ch/record/2256875
20. CMS Collaboration (2017) CMS luminosity measurements for the 2016 data taking period, CMS Physics Analysis Summary CMS-PAS-LUM-17-001, CERN, Geneva. http://cds.cern.ch/record/2257069
21. Cacciari M, Frixione S, Mangano ML, Nason P, Ridolfi G (2004) The $t\bar{t}$ cross-section at 1.8 TeV and 1.96 TeV: a study of the systematics due to parton densities and scale dependence. JHEP 04:068. arXiv:hep-ph/0303085
22. Catani S, de Florian D, Grazzini M, Nason P (2003) Soft gluon resummation for Higgs boson production at hadron colliders. JHEP 07:028. arXiv:hep-ph/0306211
23. Frederix R, Frixione S (2012) Merging meets matching in aMC@NLO. JHEP 12:061. arXiv:1209.6215
24. Skands P, Carrazza S, Rojo J (2014) Tuning PYTHIA 8.1: the Monash 2013 tune. Eur Phys J C 74:3024. arXiv:1404.5630
25. CMS Collaboration (2016) Investigations of the impact of the parton shower tuning in PYTHIA 8 in the modelling of $t\bar{t}$ at $\sqrt{s} = 8$ and 13 TeV. In: CMS CMS-PAS-TOP-16-021, CERN, Geneva. https://cds.cern.ch/record/2235192
26. Particle Data Group, Tanabashi M et al (2018) Review of particle physics. Phys Rev D 98:030001

27. Bowler MG (1981) e^+e^- production of heavy quarks in the string model. Z Phys C 11:169
28. ALEPH Collaboration (2001) Study of the fragmentation of b quarks into B mesons at the Z peak. Phys Lett B 512:30. arXiv:hep-ex/0106051
29. DELPHI Collaboration (2011) A study of the b-quark fragmentation function with the DELPHI detector at LEP I and an averaged distribution obtained at the Z pole. Eur Phys J C 71:1557. arXiv:1102.4748
30. Peterson C, Schlatter D, Schmitt I, Zerwas PM (1983) Scaling violations in inclusive e^+e^- annihilation spectra. Phys Rev D 27:105
31. Dulat S, Hou T, Gao J, Guzzi M, Huston J, Nadolsky P et al (2016) New parton distribution functions from a global analysis of quantum chromodynamics. Phys Rev D 93:033006. arXiv:1506.07443
32. CMS Collaboration (2018) Measurement of the top quark mass with lepton+jets final states using pp collisions at $\sqrt{s} = 13$TeV. Eur Phys J C 78:891. arXiv:1805.01428
33. Argyropoulos S, Sjöstrand T (2014) Effects of color reconnection on $\mathrm{t\bar{t}}$ final states at the LHC. JHEP 11:043. arXiv:1407.6653
34. Christiansen JR, Skands PZ (2015) String formation beyond leading colour. JHEP 08:003. arXiv:1505.01681
35. CMS Collaboration (2019) Measurements of $\mathrm{t\bar{t}}$ differential cross sections in proton-proton collisions at $\sqrt{s} = 13$ TeV using events containing two leptons. JHEP 02:149. arXiv:1811.06625
36. CMS Collaboration (2018) Measurement of normalized differential $\mathrm{t\bar{t}}$ cross sections in the dilepton channel from pp collisions at $\sqrt{s} = 13$ TeV. JHEP 04:060. arXiv:1708.07638
37. CMS Collaboration (2017) Measurement of differential cross sections for top quark pair production using the lepton+jets final state in proton-proton collisions at 13 TeV. Phys Rev D 95:092001. arXiv:1610.04191
38. CMS Collaboration (2016) Measurement of the top quark pair production cross section in proton-proton collisions at $\sqrt{s} = 13$ TeV. Phys Rev Lett 116:052002. arXiv:1510.05302
39. James F, Roos M (1975) Minuit: a system for function minimization and analysis of the parameter errors and correlations. Comput Phys Commun 10:343
40. Kieseler J, Illustrations provided in private communication
41. CMS Collaboration (2019) Measurement of the top quark mass in the all-jets final state at $\sqrt{s} = 13$ TeV and combination with the lepton+jets channel. Eur Phys J C 79:313. arXiv:1812.10534

Chapter 6
Extraction of the Top Quark Mass and the Strong Coupling Constant

In this chapter, the inclusive $\mathrm{t\bar{t}}$ cross section measured in Chap. 5 is used to extract the strong coupling constant and the top quark mass in the $\overline{\mathrm{MS}}$ and on-shell schemes. The strategy used for the so-called QCD analysis is outlined in Sect. 6.1, while details on the theoretical predictions used are given in Sect. 6.2. Finally, the results of the extraction of α_S and m_t are discussed in Sects. 6.3 and 6.4, respectively. All the results presented in this chapter have been obtained for the purpose of this thesis, and are documented in a paper published in the European Physics Journal [1].

6.1 The QCD Analysis

The inclusive $\mathrm{t\bar{t}}$ production cross section can be calculated in perturbative QCD, as described in Chap. 2. The result of the calculation depends on the values of the fundamental parameters of the QCD Lagrangian, in particular α_S and m_t, and on the proton PDFs. Therefore, the comparison between data and theoretical calculations can be used to experimentally determine the QCD parameters. However, α_S and m_t cannot be extracted simultaneously from a measurement of the inclusive $\sigma_{\mathrm{t\bar{t}}}$, since a change in one parameter can be compensated by an appropriate change in the other. For this reason, when one parameter is extracted, an assumption on the other one has to be made.

Theoretical predictions at a fixed order in perturbation theory are performed in a well-defined renormalization scheme, and QCD parameters extracted using these calculations can be interpreted unambiguously. This is especially important in the case of m_t, as the result of direct measurements lack a clear theoretical interpretation. However, as discussed in Chap. 5, the measured value of $\sigma_{\mathrm{t\bar{t}}}$ has an intrinsic experimental dependence on the value of m_t used in the simulation, m_t^{MC}, which affects the acceptance of the selection [2, 3]. In a QCD analysis, this dependence has to be estimated and taken into account, and an assumption in the relation between m_t^{MC} and m_t

M. M. Defranchis, *First Measurement of the Running of the Top Quark Mass*, Springer Theses, https://doi.org/10.1007/978-3-030-90376-3_6

has to be made. In particular, it is commonly assumed that $m_t^{MC} = m_t^{pole} \pm 1$ GeV [2, 3], which introduces an undesired additional uncertainty. On the opposite, the experimental dependence of the measured $\sigma_{t\bar{t}}$ on α_S is known to be negligible [2, 3].

A different approach is followed in this analysis, based the method proposed in Ref. [4]. The value of $\sigma_{t\bar{t}}$ used as input to the QCD analysis is measured simultaneously with m_t^{MC} (Chap. 5). In this way, the dependence of $\sigma_{t\bar{t}}$ on m_t^{MC} is mitigated, and no prior assumption on the relation between m_t^{MC} and m_t has to be made [4]. Furthermore, the correlation between $\sigma_{t\bar{t}}$ and m_t^{MC} is properly taken into account in the measurement, and the uncertainty in $\sigma_{t\bar{t}}$ incorporates the experimental dependence on m_t^{MC}.

6.2 Theoretical Predictions

The strong coupling constant evaluated at the mass of the Z boson, $\alpha_S(m_Z)$, is extracted using theoretical predictions in the $\overline{\text{MS}}$ scheme at NNLO, obtained using the HATHOR 2.0 software [5]. The same calculations are also used to determine the top quark mass in the $\overline{\text{MS}}$ scheme, $m_t(m_t)$, while the top quark pole mass, m_t^{pole}, is extracted using predictions in the on-shell scheme performed at NNLO+NNLL precision using the TOP++ 2.0 program [6]. In both cases, the renormalization and factorization scales are set to the value of the top quark mass.

The calculations are interfaced with the most recent available sets of NNLO PDFs: ABMP16 [7], CT14 [8], MMHT14 [9], and NNPDF3.1 [10]. Unlike in the case of the other PDFs, in ABMP16 the heavy quark masses are treated in the $\overline{\text{MS}}$ scheme. As summarized in Table 6.1, different assumptions on the QCD parameters are made in the determination of the considered PDF sets. In the MMHT14, CT14, and NNPDF3.1 PDFs, the world average value is assumed for $\alpha_S(m_Z)$, which corresponds to 0.118, while different assumptions are made on m_t^{pole}. In ABMP16, instead, $\alpha_S(m_Z)$ and $m_t(m_t)$ are determined simultaneously with the PDFs: this approach has the advantage of taking into account the correlations between the PDFs and α_S.

Table 6.1 Values of the top quark pole mass m_t^{pole} and strong coupling constant $\alpha_S(m_Z)$ used in the different PDF sets. Also shown are the corresponding $m_t(m_t)$ values obtained using the RUNDEC conversion, the number of loops in the conversion, and the α_S range used to estimate the PDF uncertainties [1]

	ABMP16	NNPDF3.1	CT14	MMHT14
m_t^{pole} (GeV)	170.37	172.5	173.3	174.2
RUNDEC loops	3	2	2	3
$m_t(m_t)$ (GeV)	160.86	162.56	163.30	163.47
$\alpha_S(m_Z)$	0.116	0.118	0.118	0.118
α_S range	0.112–0.120	0.108–0.124	0.111–0.123	0.108–0.128

Furthermore, while CT14 does not use any $t\bar{t}$ data in the PDF determination, several Tevatron and LHC results are used in the other PDFs. Although this could potentially introduce undesired correlations in the context of this analysis, it was demonstrated that the used $t\bar{t}$ data have only marginal impact on global PDF fits [7, 8].

6.3 Extraction of the Strong Coupling Constant

The value of $\alpha_S(m_Z)$ is obtained via a χ^2 fit of theoretical predictions to the measured $\sigma_{t\bar{t}}$. The fit is performed using the open-source QCD analysis framework XFITTER [11], which implements the χ^2 definition from Ref. [12]. The definition includes the uncertainty in the measured $\sigma_{t\bar{t}}$ (fit) and in the PDFs. In the HATHOR calculation, the value of $m_t(m_t)$ is set to the one utilized in the considered PDF set. The values of m_t^{pole} used in NNPDF3.1, CT14, and MMHT14 are converted to $m_t(m_t)$ using the RUNDEC code [13, 14], following the prescription provided by the corresponding group. Since the choice of $\alpha_S(m_Z)$ significantly affects the PDF determination, PDFs obtained with a series of values of $\alpha_S(m_Z)$ are provided by each PDF group. The ranges of the variation for the considered sets are listed in Table 6.1, and the value of $\alpha_S(m_Z)$ in the calculation is varied consistently with that in the PDFs.

The PDF uncertainties are estimated by using the provided PDF eigenvectors. The quoted uncertainties correspond to a 68% confidence interval, which in the case of CT14 is scaled down from the 95% confidence interval of the provided eigenvectors. For the ABMP16 and NNPDF3.1 sets, the full estimate of the PDF uncertainties is performed for each value of the $\alpha_S(m_Z)$, while the uncertainties relative to the CT14 and MMHT14 sets are only provided for the nominal PDF obtained with $\alpha_S(m_Z) = 0.118$. In this case, the uncertainty of the nominal PDF is used for all values of $\alpha_S(m_Z)$. The resulting χ^2 profiles as a function of $\alpha_S(m_Z)$ are shown in Fig. 6.1 (left) for the considered PDFs.

The best-fit value of $\alpha_S(m_Z)$, indicated with α_S^{min}, is extracted by parametrizing the χ^2 profiles with the function:

$$\chi^2(\alpha_S) = \chi^2_{min} + \left(\frac{\alpha_S - \alpha_S^{min}}{\delta(\alpha_S^{min})}\right)^2, \tag{6.1}$$

where $\delta(\alpha_S^{min})$ represents the combined fit+PDF uncertainty in α_S^{min}. In Fig. 6.1 (right), the results obtained for the different PDF sets are compared to the 2018 PDG world average [15]. The uncertainty due to missing higher order corrections is estimated by varying the renormalization and factorization scales in the calculation by a factor of two up and down, avoiding cases in which $\mu_r/\mu_f = 4$ or 1/4. The difference between the envelope of the resulting values of $\alpha_S(m_Z)$ and its nominal best-fit value is taken as the corresponding scale uncertainty. The values of the extracted $\alpha_S(m_Z)$ and their uncertainties are summarized in Table 6.2. The results are found to

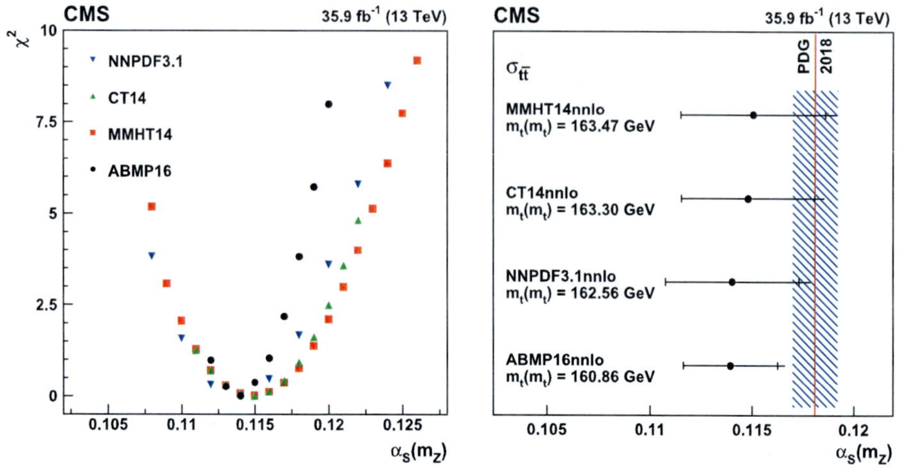

Fig. 6.1 Left: χ^2 versus α_S obtained from the comparison of the measured $\sigma_{t\bar{t}}$ value to the NNLO prediction in the $\overline{\mathrm{MS}}$ scheme using different PDFs (symbols of different styles). Right: $\alpha_S(m_Z)$ obtained from the comparison of the measured $\sigma_{t\bar{t}}$ value to the theoretical prediction using different PDF sets in the $\overline{\mathrm{MS}}$ scheme. The corresponding value of $m_t(m_t)$ is given for each PDF set. The inner horizontal bars on the points represent the experimental and PDF uncertainties added in quadrature. The outer horizontal bars show the total uncertainties. The vertical line displays the world-average $\alpha_S(m_Z)$ value, with the hatched band representing its uncertainty [1]

Table 6.2 Values of $\alpha_S(m_Z)$ with their uncertainties obtained from a comparison of the measured $\sigma_{t\bar{t}}$ value to the NNLO prediction in the $\overline{\mathrm{MS}}$ scheme using different PDF sets. The first uncertainty is the combination of the experimental and PDF uncertainties, and the second is from the variation of the renormalization and factorization scales [1]

PDF set	$\alpha_S(m_Z)$
ABMP16	0.1139 ± 0.0023 (fit + PDF) $^{+0.0014}_{-0.0001}$ (scale)
NNPDF3.1	0.1140 ± 0.0033 (fit + PDF) $^{+0.0021}_{-0.0002}$ (scale)
CT14	0.1148 ± 0.0032 (fit + PDF) $^{+0.0018}_{-0.0002}$ (scale)
MMHT14	0.1151 ± 0.0035 (fit + PDF) $^{+0.0020}_{-0.0002}$ (scale)

be in good agreement with each other, and represent the most precise determination of $\alpha_S(m_Z)$ at a hadron collider, to date. It can be noted that the most precise result is obtained with the ABMP16 PDF set. This can be attributed to the fact that the correlation between α_S and the gluon PDF is to large extent taken into account in the determination of this PDF.

The dependence of the extracted values of $\alpha_S(m_Z)$ on the assumed value of $m_t(m_t)$ in the calculation is shown in Fig. 6.2 for the considered PDFs. In this case, the value of $m_t(m_t)$ is varied only in the calculation. This dependence is found to be less pronounced in case of the ABMP16 set, due to a partial decorrelation of the two QCD parameters through their simultaneous extraction in the ABMP16 fit.

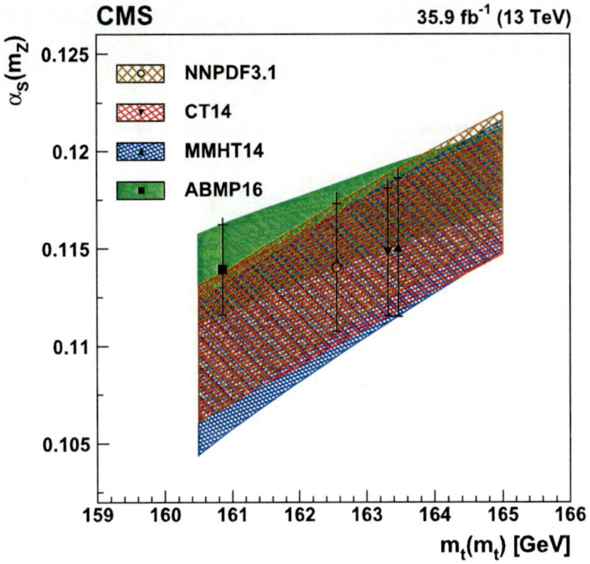

Fig. 6.2 Values of $\alpha_S(m_Z)$ obtained in the comparison of the $\sigma_{t\bar{t}}$ measurement to the NNLO prediction using different PDFs, as a function of the $m_t(m_t)$ value used in the theoretical calculation. The results from using different PDFs are shown by the bands with different shadings, with the band width corresponding to the quadratic sum of the experimental and PDF uncertainties in $\alpha_S(m_Z)$. The resulting measured values of $\alpha_S(m_Z)$ are shown by the different style points at the $m_t(m_t)$ values used for each PDF. The inner vertical bars on the points represent the quadratic sum of the experimental and PDF uncertainties in $\alpha_S(m_Z)$, while the outer vertical bars show the total uncertainties [1]

Finally, the values of $\alpha_S(m_Z)$ obtained with the different choices of the renormalization and factorization scales are shown in Fig. 6.3 for the considered PDFs. The result is found to be sensitive to the choice of the renormalization scale, while the dependence on the factorization scale is found to be negligible. For illustration, Fig. 6.4 compares the scale dependence of the values of $\alpha_S(m_Z)$ extracted using NNLO calculations in the $\overline{\text{MS}}$ and on-shell schemes, performed with the HATHOR software interfaced to the ABMP16 PDFs. A significantly larger effect is observed in the latter case, which reflects the faster convergence of the perturbative series in the calculation using the top quark mass defined in the $\overline{\text{MS}}$ scheme, as discussed in Chap. 2.

6.4 Extraction of the Top Quark Mass

The same procedure is also used to extract the value of $m_t(m_t)$ using different PDFs. In this case, the value of $\alpha_S(m_Z)$ in the calculation is set to that of the considered PDF set, and the χ^2 profiles are scanned as a function of $m_t(m_t)$ in steps of 0.5 GeV.

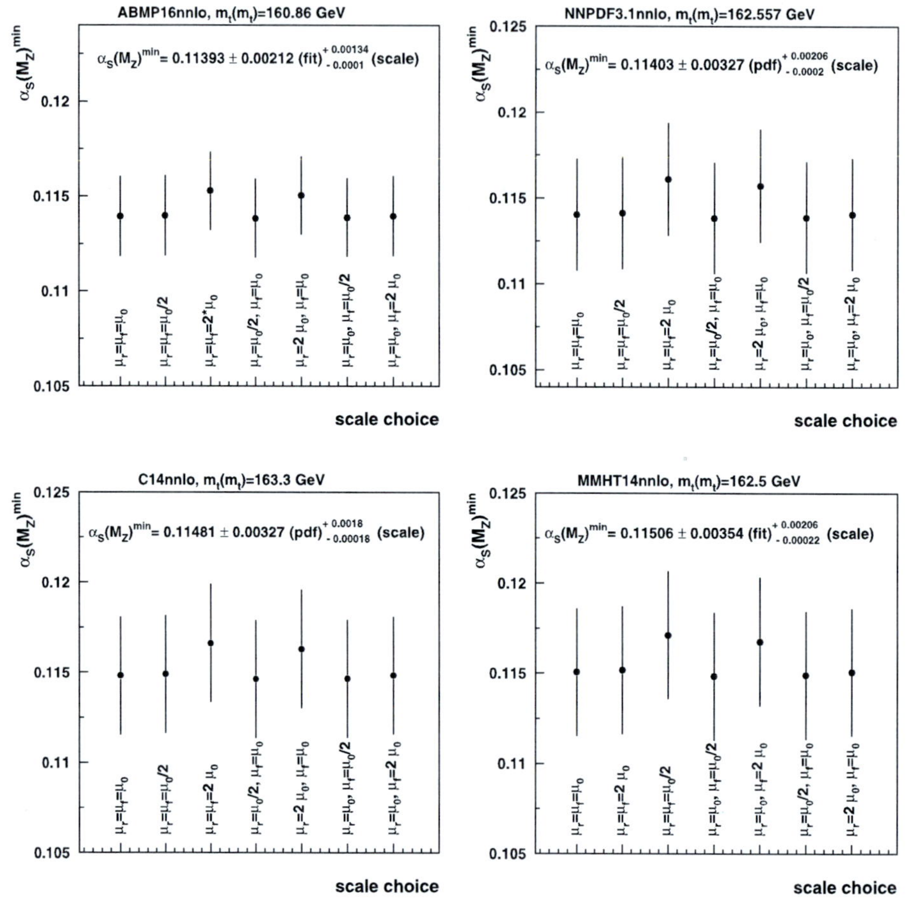

Fig. 6.3 Extracted value of $\alpha_S(m_Z)$ as a function of the choice of the renormalization and factorization scales, using the ABMP16 (upper left), NNPDF3.1 (upper right), CT14 (lower left), and MMHT14 (lower right) PDF sets

In the case of ABMP16, $m_t(m_t)$ is varied between 158 and 163 GeV, while it is varied between 162 and 167 GeV for the other PDFs. Since the impact of the choice of $m_t(m_t)$ on global PDF fits is not expected to be significant, the value of $m_t(m_t)$ is not varied in PDFs. In addition to the uncertainties described in the previous section, the uncertainty in the value of $\alpha_S(m_Z)$ is propagated to $m_t(m_t)$. In the case of ABMP16, where α_S is determined simultaneously with the PDFs, this uncertainty is incorporated in the PDF eigenvectors. For the other sets, instead, the uncertainty is evaluated by using alternative eigenvectors where the value of $\alpha_S(m_Z)$ is varied by ± 0.001, which corresponds to the uncertainty in the world-average value [15].

The χ^2 profiles obtained with the considered PDF sets are shown in Fig. 6.5, and the extracted values of $m_t(m_t)$ are shown in Fig. 6.6 and summarized in Table 6.3. These results represent the most precise determination of $m_t(m_t)$, to date. It can be noted that, although the uncertainties are of similar size, the value of $m_t(m_t)$ obtained

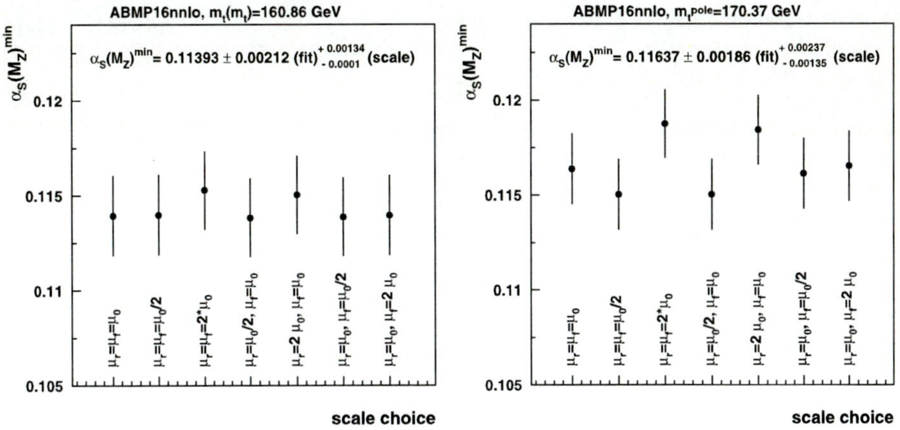

Fig. 6.4 Extracted value of $\alpha_S(m_Z)$ as a function of the choice of the renormalization and factorization scales in HATHOR NNLO calculation in the $\overline{\text{MS}}$ (left) and on-shell (right) renormalization schemes, interfaced to the ABMP16 PDF set

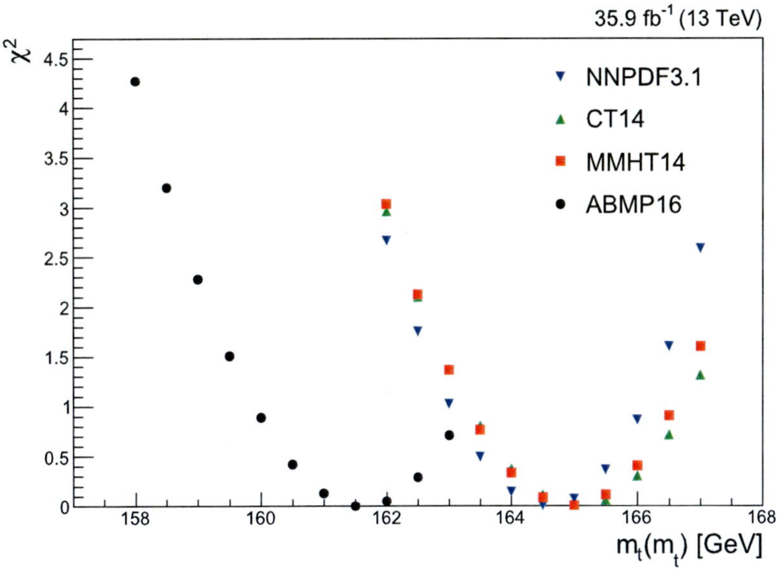

Fig. 6.5 Data-to-theory χ^2 as a function of $m_t(m_t)$ for different sets of PDFs

with ABMP16 is lower than the others. This is due to the fact that the value of $\alpha_S(m_Z)$ determined in the ABMP16 fit, and therefore used in the calculation, is lower than the one used in the other PDF sets (Table 6.1). The lower value of $\alpha_S(m_Z)$ is therefore compensated by a lower value of $m_t(m_t)$, as expected from Fig. 6.2. This effect can be mitigated by performing a simultaneous determination of m_t and α_S, as done, e.g. in Ref. [16].

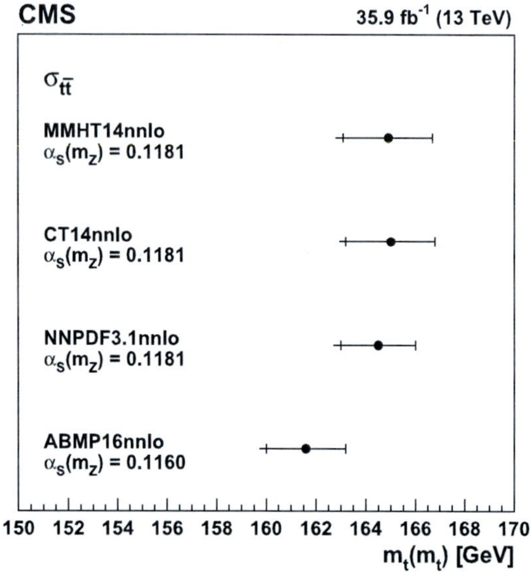

Fig. 6.6 Values of $m_{\mathrm{t}}(m_{\mathrm{t}})$ obtained from comparing the $\sigma_{\mathrm{t\bar{t}}}$ measurement to the theoretical NNLO predictions using different PDF sets. The inner horizontal bars on the points represent the quadratic sum of the experimental, PDF, and $\alpha_S(m_Z)$ uncertainties, while the outer horizontal bars give the total uncertainty [1]

Table 6.3 Values of $m_{\mathrm{t}}(m_{\mathrm{t}})$ obtained from the comparison of the $\sigma_{\mathrm{t\bar{t}}}$ measurement with the NNLO predictions using different PDF sets. The first uncertainty shown comes from the experimental, PDF, and $\alpha_S(m_Z)$ uncertainties, and the second from the variation in the renormalization and factorization scales [1]

PDF set	$m_{\mathrm{t}}(m_{\mathrm{t}})$ (GeV)
ABMP16	161.6 ± 1.6 (fit + PDF + α_S) $^{+0.1}_{-1.0}$ (scale)
NNPDF3.1	164.5 ± 1.6 (fit + PDF + α_S) $^{+0.1}_{-1.0}$ (scale)
CT14	165.0 ± 1.8 (fit + PDF + α_S) $^{+0.1}_{-1.0}$ (scale)
MMHT14	164.9 ± 1.8 (fit + PDF + α_S) $^{+0.1}_{-1.1}$ (scale)

The values of $m_{\mathrm{t}}(m_{\mathrm{t}})$ obtained with different choices for the renormalization and factorization scales are shown in Fig. 6.7. As in the case of α_S, the dependence on the factorization scale is found to be small, and the scale uncertainty is mainly determined by the variation of the renormalization scale.

Finally, the value of $m_{\mathrm{t}}^{\mathrm{pole}}$ is determined using theoretical predictions obtained with the TOP++ program, following the procedure used for the extraction of $m_{\mathrm{t}}(m_{\mathrm{t}})$. The resulting values of $m_{\mathrm{t}}^{\mathrm{pole}}$ and their uncertainties are summarized in Table 6.4 for the considered sets of PDFs. It can be noted that, despite the inclusion of NNLL corrections, the impact of scale uncertainties is larger than in the case of $m_{\mathrm{t}}(m_{\mathrm{t}})$. This demonstrates once more the faster perturbative convergence of the calculations performed in the $\overline{\mathrm{MS}}$ scheme.

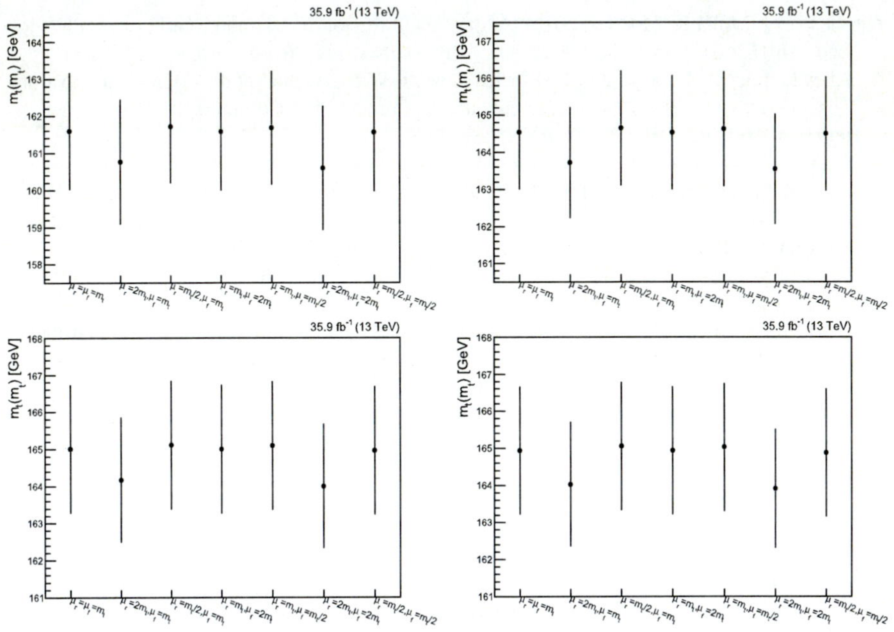

Fig. 6.7 Extracted value of $m_{\mathrm{t}}(m_{\mathrm{t}})$ as a function of the choice of the renormalization and factorization scales, using the ABMP16 (upper left), NNPDF3.1 (upper right), CT14 (lower left), and MMHT14 (lower right) PDF sets

Table 6.4 Values of $m_{\mathrm{t}}^{\mathrm{pole}}$ obtained by comparing the $\sigma_{\mathrm{t\bar{t}}}$ measurement with predictions at NNLO+NNLL using different PDF sets [1]

PDF set	$m_{\mathrm{t}}^{\mathrm{pole}}$ (GeV)
ABMP16	169.9 ± 1.8 (fit + PDF + α_S) $^{+0.8}_{-1.2}$ (scale)
NNPDF3.1	173.2 ± 1.9 (fit + PDF + α_S) $^{+0.9}_{-1.3}$ (scale)
CT14	173.7 ± 2.0 (fit + PDF + α_S) $^{+0.9}_{-1.4}$ (scale)
MMHT14	173.6 ± 1.9 (fit + PDF + α_S) $^{+0.9}_{-1.4}$ (scale)

References

1. *CMS* Collaboration (2019) Measurement of the $\mathrm{t\bar{t}}$ production cross section, the top quark mass, and the strong coupling constant using dilepton events in pp collisions at $\sqrt{s} = 13$TeV. Eur Phys J C 79:368. https://doi.org/10.1140/epjc/s10052-019-6863-8. arXiv:1812.10505
2. *CMS* Collaboration (2014) Determination of the top-quark pole mass and strong coupling constant from the $\mathrm{t\bar{t}}$ production cross section in pp collisions at $\sqrt{s} = 7$TeV. Phys Lett B 728:496. https://doi.org/10.1016/j.physletb.2014.08.040, https://doi.org/10.1016/j.physletb.2013.12.009. arXiv:1307.1907
3. *CMS* Collaboration (2016) Measurement of the $\mathrm{t\bar{t}}$ production cross section in the $e\mu$ channel in proton-proton collisions at $\sqrt{s} = 7$ and 8 TeV. JHEP 08:029. https://doi.org/10.1007/JHEP08(2016)029. arXiv:1603.02303

4. Kieseler J, Lipka K, Moch S (2016) Calibration of the top-quark monte carlo mass. Phys Rev Lett 116:162001. https://doi.org/10.1103/PhysRevLett.116.162001. arXiv:1511.00841
5. Aliev M, Lacker H, Langenfeld U, Moch S, Uwer P, Wiedermann M (2011) HATHOR: HAdronic Top and Heavy quarks crOss section calculatoR. Comput Phys Commun 182:1034. https://doi.org/10.1016/j.cpc.2010.12.040. arXiv:1007.1327
6. Czakon M, Mitov A (2014) TOP++: a program for the calculation of the top-pair cross-section at hadron colliders. Comput Phys Commun 185:2930. https://doi.org/10.1016/j.cpc.2014.06.021. arXiv:1112.5675
7. Alekhin S, Blümlein J, Moch S, Placakyte R (2017) Parton distribution functions, α_S, and heavy-quark masses for LHC Run II. Phys Rev D 96:014011. https://doi.org/10.1103/PhysRevD.96.014011. arXiv:1701.05838
8. Dulat S, Hou T, Gao J, Guzzi M, Huston J, Nadolsky P et al (2016) New parton distribution functions from a global analysis of quantum chromodynamics. Phys Rev D 93:033006. https://doi.org/10.1103/PhysRevD.93.033006. arXiv:1506.07443
9. Harland-Lang LA, Martin AD, Motylinski P, Thorne RS (2015) Parton distributions in the LHC era: MMHT 2014 PDFs. Eur Phys J C 75:204. https://doi.org/10.1140/epjc/s10052-015-3397-6. arXiv:1412.3989
10. *NNPDF* Collaboration (2017) Parton distributions from high-precision collider data. Eur Phys J C 77:663. https://doi.org/10.1140/epjc/s10052-017-5199-5. arXiv:1706.00428
11. Alekhin S et al (2015) HERAFitter. Eur Phys J C 75:304. https://doi.org/10.1140/epjc/s10052-015-3480-z. arXiv:1410.4412
12. H1 and ZEUS Collaborations (2015) Combination of measurements of inclusive deep inelastic $e^{\pm}p$ scattering cross sections and QCD analysis of HERA data. Eur Phys J C 75:580. https://doi.org/10.1140/epjc/s10052-015-3710-4. arXiv:1506.06042
13. Chetyrkin KG, Kühn JH, Steinhauser M (2000) RUNDEC: a Mathematica package for running and decoupling of the strong coupling and quark masses. Comput Phys Commun 133:43. https://doi.org/10.1016/S0010-4655(00)00155-7. arXiv:hep-ph/0004189
14. Schmidt B, Steinhauser M (2012) CRUNDEC: a C++ package for running and decoupling of the strong coupling and quark masses. Comput Phys Commun 183:1845. https://doi.org/10.1016/j.cpc.2012.03.023. arXiv:1201.6149
15. Particle Data Group, Tanabashi M, et al (2018) Review of particle physics. Phys Rev D 98:030001 https://doi.org/10.1103/PhysRevD.98.030001
16. *CMS* Collaboration (2019) Measurement of $t\bar{t}$ normalised multi-differential cross sections in pp collisions at $\sqrt{s} = 13$ TeV, and simultaneous determination of the strong coupling strength, top quark pole mass, and parton distribution functions. (Submitted to) Eur Phys J. arXiv:1904.05237

Chapter 7
Measurement of the Running of the Top Quark Mass

The first measurement of the running of the top quark mass, developed and performed in the scope of this thesis, is presented in this chapter. This measurement represents an experimental investigation of the validity of perturbative QCD up to scales of the order of 1 TeV, and is documented in a paper published in the Physics Letters journal [1]. The chapter is structured as follows: the analysis strategy is discussed in Sect. 7.1, the measurement of the differential $\mathrm{t\bar{t}}$ cross section used for the QCD interpretation is described in Sect. 7.2, and the extraction of the top quark mass running in the $\overline{\mathrm{MS}}$ scheme is presented in Sect. 7.3.

7.1 Analysis Strategy

Beyond leading order in perturbation theory, the parameters of the QCD Lagrangian undergo renormalization. However, as discussed in Chap. 2, the choice of the renormalization scheme is not unique. In the $\overline{\mathrm{MS}}$ scheme, the strong coupling constant and the quark masses depend on the renormalization scale. The scale evolution, or running, of α_S and m_t is described by the renormalization group equations (RGEs) of Eqs. 2.14 and 2.15, respectively. The RGEs can be solved in perturbation theory by expanding the β function and the anomalous mass dimension in powers of α_S. At one-loop precision, an analytical expression for the running of the top quark mass can be obtained:

$$m_t(\mu) = m_t(\mu_0)\left[1 - \frac{1}{\pi}\alpha_S(\mu)\ln\left(\frac{\mu^2}{\mu_0^2}\right)\right]. \tag{7.1}$$

Here, μ_0 is the initial scale at which m_t is evaluated, μ denotes an arbitrarily chosen renormalization scale, and $m_t(\mu)$ is the $\overline{\mathrm{MS}}$ mass of the top quark evaluated at the scale μ. In order to ensure the validity of perturbation theory, the scales μ and μ_0 are required to be significantly larger than Λ_{QCD}.

M. M. Defranchis, *First Measurement of the Running of the Top Quark Mass*, Springer Theses, https://doi.org/10.1007/978-3-030-90376-3_7

In a theoretical calculation, the renormalization scale is usually identified with the physical scale of the process Q. Therefore, the running of a QCD parameter, e.g. α_S, can be experimentally determined by extracting the value of α_S from a sensitive observable measured as a function of the scale Q. From a practical point of view, α_S can be determined at a fixed reference scale, e.g. $\mu = m_Z$, and then evolved to the scale Q using the corresponding RGE [2]. Since the RGEs are implicitly assumed in theoretical calculations, this approach is equivalent to determining $\alpha_S(Q)$ directly, e.g. as in Ref. [3]. Furthermore, the order at which the running of a QCD parameter is determined depends on the order of the calculation.

The same procedure can be used to extract the running of the mass of a heavy quark, e.g. as described in Ref. [4] for the charm quark mass. While in the case of α_S the reference scale is often chosen as the mass of the Z boson, the quark masses are usually evaluated at the scale of the quark mass itself.

In this analysis, the running of m_t in the $\overline{\text{MS}}$ scheme is determined at one-loop precision from a measurement of the differential $\text{t}\bar{\text{t}}$ cross section as a function of the invariant mass of the $\text{t}\bar{\text{t}}$ system, $m_{\text{t}\bar{\text{t}}}$. The top quark mass is extracted as a function of $m_{\text{t}\bar{\text{t}}}$ using NLO theoretical predictions of $\text{d}\sigma_{\text{t}\bar{\text{t}}}/\text{d}m_{\text{t}\bar{\text{t}}}$ where the heavy quark masses are treated in the $\overline{\text{MS}}$ scheme [5]. In order to compare the result of the measurement to fixed-order calculations, the differential cross section is determined at the parton level, before parton-shower radiation.

7.2 Measurement of the Differential Cross Section

The measurement of $\text{d}\sigma_{\text{t}\bar{\text{t}}}/\text{d}m_{\text{t}\bar{\text{t}}}$ is performed by extending the method presented in Chap. 5. The analysis also makes use of the same data set, MC samples, and event selection as the one presented in Chap. 5, and the same sources of systematic uncertainties are considered. In this case, however, the uncertainty in the integrated luminosity is profiled in the likelihood together with the other systematic uncertainties.

In addition, in order to reconstruct the invariant mass of the $\text{t}\bar{\text{t}}$ system, $m_{\text{t}\bar{\text{t}}}^{\text{reco}}$, an analytical reconstruction of the top quark and antiquark kinematic variables is performed in events with at least two jets, using the algorithm described in Refs. [6, 7]. The goal of the kinematic reconstruction is to infer the momenta of the neutrinos originating from the decays of the W bosons using information about the reconstructed final-state objects and the missing transverse energy. The algorithm explores all possible combinations of leptons and jets and solves a system of equations that relate the four-momenta of the decay products of the $\text{t}\bar{\text{t}}$ pair under the following assumptions: the missing transverse energy arises uniquely from the two neutrinos originating from the decays of the W bosons; the invariant mass of the reconstructed decay products of each W boson corresponds to the W boson mass, which is assumed to be 80.4 GeV; and the invariant mass of the decay products of the top quark and antiquark corresponds to a predefined value of the top quark mass, denoted with $m_{\text{t}}^{\text{kin}}$. In this way, the eight unknown variables related to the neutrino momenta are constrained by four kinematic assumptions, leaving the kinematic equations with a four-fold ambiguity. In this analysis, m_t^{kin} is set to the value of the top quark mass used in the simulation, m_t^{MC}, which in the nominal $t\bar{t}$ sample corresponds to 172.5 GeV. In order to account for detector effects, solutions are looked for by smearing the reconstructed momenta of the leptons and jets according to their resolution. Similarly, the width of the W boson is accounted for by smearing the mass of the W boson according to a Breit–Wigner distribution. Over 100 smearings are performed for each considered combination of leptons and jets. In order to maximize the correct assignments, the solution that yields the smallest value of $m_{t\bar{t}}$ is chosen [8]. A weight is assigned to each selected solution based on the expected spectrum of the invariant mass of the lepton and jet originating from the same top quark. For each combination, a total weight is defined as the sum of the weights assigned to the individual smearings. If different possible combinations are found, the one that includes the largest number of b-tagged jets is chosen, and if more than one solution with the same number of b-tagged jets are found, the one with the highest total weight is selected. For the chosen combination, the kinematic properties of the $t\bar{t}$ system are estimated as the weighted average of the solutions obtained in the various smearings. The efficiency of the kinematic reconstruction is found to be about 90% [7], while the resolution in $m_{t\bar{t}}^{reco}$ is estimated to be about 13%.

The kinematic reconstruction introduces an additional undesired dependence on m_t^{kin}. In this analysis, the effect is estimated by repeating the event selection and the kinematic reconstruction with three different values of m_t^{kin}, corresponding to 169.5, 172.5, and 175.5 GeV, and the value of m_t^{MC} is varied accordingly. In this analysis, therefore, m_t^{kin} and m_t^{MC} are considered the same parameter. As in the measurement described in Chap. 5, the dependence on the top quark mass is fully incorporated in the fit, where m_t^{MC} is treated as a free parameter.

The differential cross section is determined directly at the parton level by means of a maximum-likelihood fit to distributions of final-state observables. This method is known as maximum-likelihood unfolding and, unlike other unfolding methods, allows the systematic uncertainties to be constrained together with the parameters of interest. In order to perform the unfolding, the $t\bar{t}$ sample is split into subsamples in intervals of parton-level $m_{t\bar{t}}$ corresponding to the desired bins in $d\sigma_{t\bar{t}}/dm_{t\bar{t}}$. Each subsample is then treated as an independent signal, and represents the $t\bar{t}$ production cross section at the physical scale μ_k, defined as the centre-of-gravity of bin k in $m_{t\bar{t}}$. However, the results presented in Sect. 7.3 do not depend on the exact choice of μ_k, provided that it represents the physical energy scale of the process in that bin. In the figures, the subprocess corresponding to bin k is indicated with "Signal (μ_k)". The bin boundaries, the fraction of events in each bin, and the scales μ_k as determined using the nominal POWHEG simulation are summarized in Table 7.1. The bin widths $\Delta m_{t\bar{t}}^k$ are chosen taking into account the experimental resolution in $m_{t\bar{t}}$. The distribution of $m_{t\bar{t}}^{reco}$ after the fit to the data, obtained with the binning of Table 7.1, is compared to the data in Fig. 7.1. The same distribution with finer binning can be found in Fig. 7.2, where the post-fit distributions of the p_T of the reconstructed top quark, antiquark, and $t\bar{t}$ system are also shown.

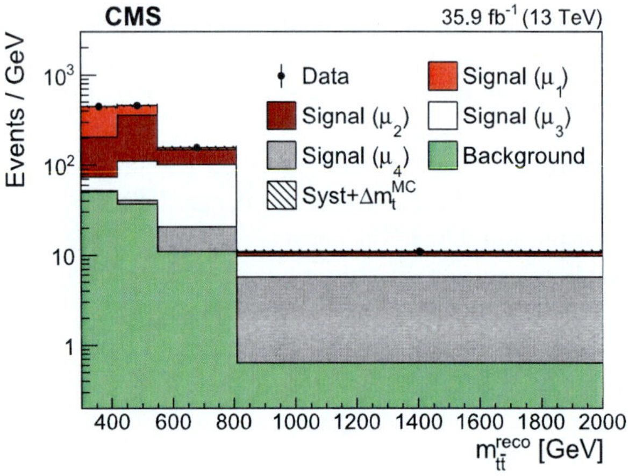

Fig. 7.1 Distribution of $m_{t\bar{t}}^{\text{reco}}$ after the fit to the data, with the same binning as used in the fit. The hatched band corresponds to the total uncertainty in the predicted yields, including the contribution from m_{t}^{MC} ($\Delta m_{\text{t}}^{\text{MC}}$) and all correlations. The $\text{t}\bar{\text{t}}$ MC sample is split into four subsamples, denoted with "Signal (μ_k)", corresponding to bins of $m_{t\bar{t}}$ at the parton level. The first and last bins contain all events with $m_{t\bar{t}}^{\text{reco}} < 420\,\text{GeV}$ and $m_{t\bar{t}}^{\text{reco}} > 810\,\text{GeV}$, respectively [1]

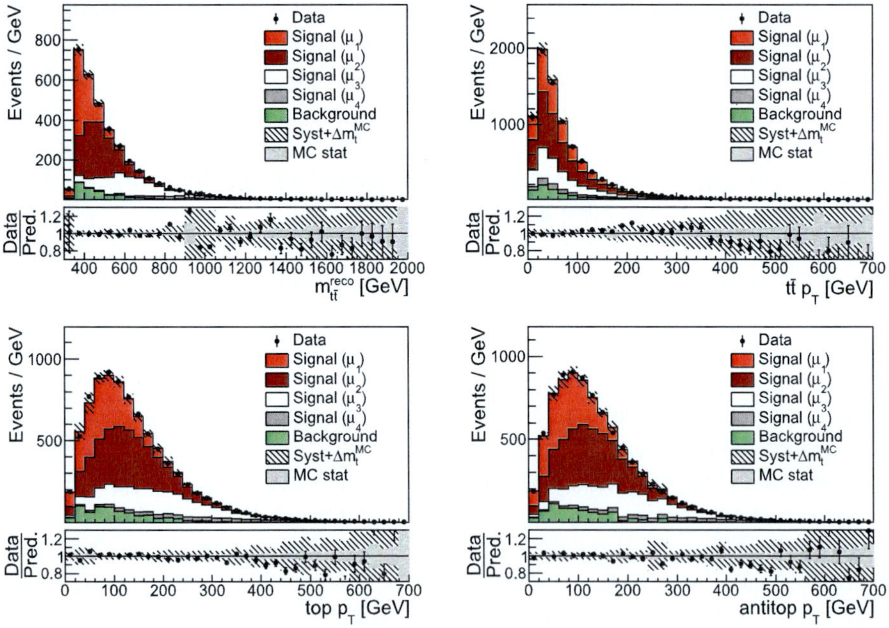

Fig. 7.2 Distributions of the reconstructed invariant mass (upper left) and transverse momentum (upper right) of the $\text{t}\bar{\text{t}}$ system, and of the p_{T} of the top quark (lower left) and antiquark (lower right) after the fit to the data. The details of the plots are the same as in Fig. 7.1

Table 7.1 Bins of $m_{\mathrm{t\bar{t}}}$, the corresponding fraction of events in the POWHEG simulation, and the representative scale μ_k [1]

Bin	$m_{\mathrm{t\bar{t}}}$ [GeV]	Fraction [%]	μ_k [GeV]
1	<420	30	384
2	420–550	39	476
3	550–810	24	644
4	>810	7	1024

The same likelihood function as the one defined in Chap. 5 is used. In this case, however, the expected number of events ν_i can be expressed as:

$$\nu_i = \sum_{k=1}^{4} s_i^k(\sigma_{\mathrm{t\bar{t}}}^{(\mu_k)}, m_{\mathrm{t}}^{\mathrm{MC}}, \vec{\lambda}) + \sum_j b_i^j(m_{\mathrm{t}}^{\mathrm{MC}}, \vec{\lambda}, \omega_j)\,. \tag{7.2}$$

Here, s_i^k is the expected number of signal events in bin k, b_i^j is the expected number of background events from source j, $\sigma_{\mathrm{t\bar{t}}}^{(\mu_k)} = (\mathrm{d}\sigma_{\mathrm{t\bar{t}}}/\mathrm{d}m_{\mathrm{t\bar{t}}})\Delta m_{\mathrm{t\bar{t}}}^k$ is the integrated cross section in bin k of $m_{\mathrm{t\bar{t}}}$, and $\vec{\lambda}$ and ω_j represent the nuisance and background normalization parameters, respectively. Due to the kinematic reconstruction, the background and data distributions also depend on $m_{\mathrm{t}}^{\mathrm{MC}}$. Since Eq. 7.2 relates the parton-level cross sections $\sigma_{\mathrm{t\bar{t}}}^{(\mu_k)}$ to the observed number of events in each reconstructed distribution, the response matrix and its dependence on the fit parameters are incorporated in the likelihood function, and the maximization of the likelihood provides results that are automatically unfolded to the parton level. Having chosen a very coarse binning in $m_{\mathrm{t\bar{t}}}$, the unfolding problem is found to be well conditioned, and no regularization is necessary. In particular, the condition number of the response matrix is estimated to be about 3.

Events are categorized according to their b-tagged jet multiplicities, and the number of events in each category is parametrized following the method described in Chap. 5. In this case, the parametrization is derived for each bin k of $m_{\mathrm{t\bar{t}}}$ independently:

$$\begin{aligned} S_{1\mathrm{b}}^k &= \mathcal{L}\sigma_{\mathrm{t\bar{t}}}^{(\mu_k)} A_{\ell\ell}^k \epsilon_{\ell\ell}^k 2\epsilon_{\mathrm{b}}^k (1 - C_{\mathrm{b}}^k \epsilon_{\mathrm{b}}^k), \\ S_{2\mathrm{b}}^k &= \mathcal{L}\sigma_{\mathrm{t\bar{t}}}^{(\mu_k)} A_{\ell\ell}^k \epsilon_{\ell\ell}^k C_{\mathrm{b}}^k (\epsilon_{\mathrm{b}}^k)^2, \\ S_{\mathrm{other}}^k &= \mathcal{L}\sigma_{\mathrm{t\bar{t}}}^{(\mu_k)} A_{\ell\ell}^k \epsilon_{\ell\ell}^k \left[1 - 2\epsilon_{\mathrm{b}}^k (1 - C_{\mathrm{b}}^k \epsilon_{\mathrm{b}}^k) - C_{\mathrm{b}}^k (\epsilon_{\mathrm{b}}^k)^2\right]. \end{aligned} \tag{7.3}$$

In order to effectively constrain each individual $\sigma_{\mathrm{t\bar{t}}}$, events with at least two jets are split in subcategories of $m_{\mathrm{t\bar{t}}}^{\mathrm{reco}}$, using the same binning as for $m_{\mathrm{t\bar{t}}}$. Events with less than two jets are instead assigned to separate subcategories. In each subcategory, a suitable observable is chosen as input to the fit. The event yield is fitted in events with less than two jets, in the last bin of $m_{\mathrm{t\bar{t}}}^{\mathrm{reco}}$, and in subcategories with zero or more than two b-tagged jets. Depending on the category, this choice is made either

Table 7.2 Input distributions to the fit in the different event categories. The number of jets, the number of b-tagged jets, the number of events, and the p_T of the softest jet are denoted with N_{jets}, N_b, N_{events}, and "jet p_T^{min}", respectively, while the category corresponding to bin k in $m_{t\bar{t}}^{reco}$ is indicated with "$m_{t\bar{t}}^{reco}$ k" [1]

	$N_b = 1$	$N_b = 2$	other N_b
$N_{jets} < 2$	N_{events}	n.a.	N_{events}
$m_{t\bar{t}}^{reco}$ 1	$m_{\ell b}^{min}$	jet p_T^{min}	N_{events}
$m_{t\bar{t}}^{reco}$ 2	$m_{\ell b}^{min}$	jet p_T^{min}	N_{events}
$m_{t\bar{t}}^{reco}$ 3	$m_{\ell b}^{min}$	jet p_T^{min}	N_{events}
$m_{t\bar{t}}^{reco}$ 4	N_{events}	N_{events}	N_{events}

to mitigate the sensitivity to the backgrounds or to reduce the impact of the statistical uncertainty of the simulation. In other categories with one b-tagged jet, the $m_{\ell b}^{min}$ distribution is used to constrain m_t^{MC}, while the p_T of the softest reconstructed jet is used in remaining categories with two b-tagged jets in order to constrain the jet energy scale uncertainties. The choice of the input distributions is summarized in Table 7.2.

The efficiency of the kinematic reconstruction in the simulation is corrected in order to match the one observed in data. In a dedicated study, the difference between the efficiency in data and MC was found be about 0.2% [7]. An uncertainty of 0.2% in the kinematic reconstruction efficiency is assigned to each signal independently. The same uncertainty is assigned to all $t\bar{t}$ events with one or two b-tagged jets, independently, while an uncertainty of 0.5% is conservatively assigned to other $t\bar{t}$ events. This approach allows any small dependence of the kinematic reconstruction on the event topology or on the number of b-tagged jets to be taken into account in the fit. Furthermore, an additional uncertainty of 1% is assigned to the sum of the background contributions in each category of $m_{t\bar{t}}^{reco}$, independently.

The input distributions after the fit are compared to the data in Fig. 7.3, and good agreement is observed. The pulls and constraints of the nuisance parameters corresponding to the modelling and all other uncertainties are shown in Fig. 7.4 and 7.5, respectively. No pull significantly beyond one standard deviation is observed, and moderate constraints on the nuisance parameters are obtained.

In the fit, the values of $\sigma_{t\bar{t}}^{(\mu_k)}$ are measured to be:

$$\sigma_{t\bar{t}}^{(\mu_1)} = 255 \pm 11\,(\text{syst}) \pm 2\,(\text{stat})\ \text{pb},$$
$$\sigma_{t\bar{t}}^{(\mu_2)} = 315 \pm 15\,(\text{syst}) \pm 2\,(\text{stat})\ \text{pb},$$
$$\sigma_{t\bar{t}}^{(\mu_3)} = 181 \pm 9\,(\text{syst}) \pm 1\,(\text{stat})\ \text{pb},$$
$$\sigma_{t\bar{t}}^{(\mu_4)} = 50 \pm 3\,(\text{syst}) \pm 1\,(\text{stat})\ \text{pb},$$

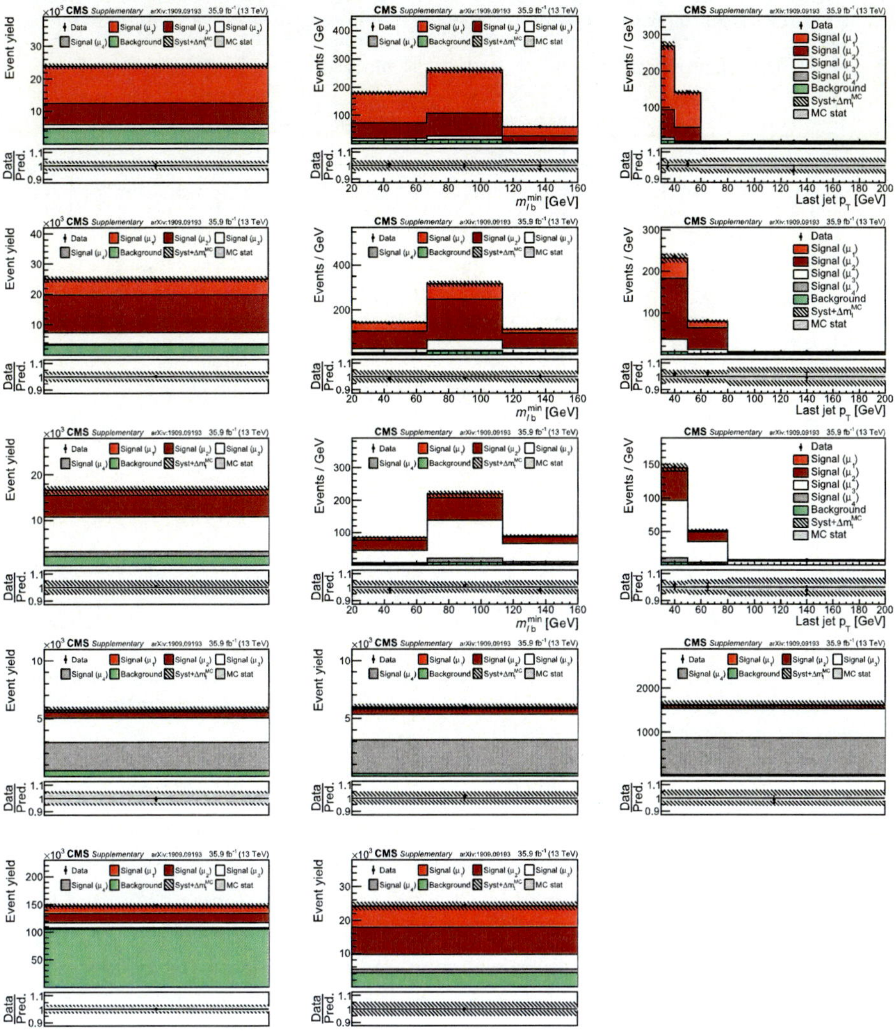

Fig. 7.3 Comparison between data (points) and post-fit distributions of the expected signal and backgrounds from simulation (shaded histograms) used in the fit of $\mathrm{d}\sigma_{t\bar{t}}/\mathrm{d}m_{t\bar{t}}$. In the left column events with zero or three or more b-tagged jets are shown. The middle (right) column shows events with exactly one (two) b-tagged jets. Events in the first, second, third, and fourth bin of $m_{t\bar{t}}^{\mathrm{reco}}$ are shown in the first, second, third, and fourth row, respectively, while events with less than two jets are shown in the fifth row. The hatched bands correspond to the total uncertainty in the sum of the predicted yields and include the contribution from the top quark mass (Δm_t^{MC}). The ratios of data to the sum of the predicted yields are shown in the lower panel of each figure. Here, the solid grey band represents the contribution of the statistical uncertainty [1].

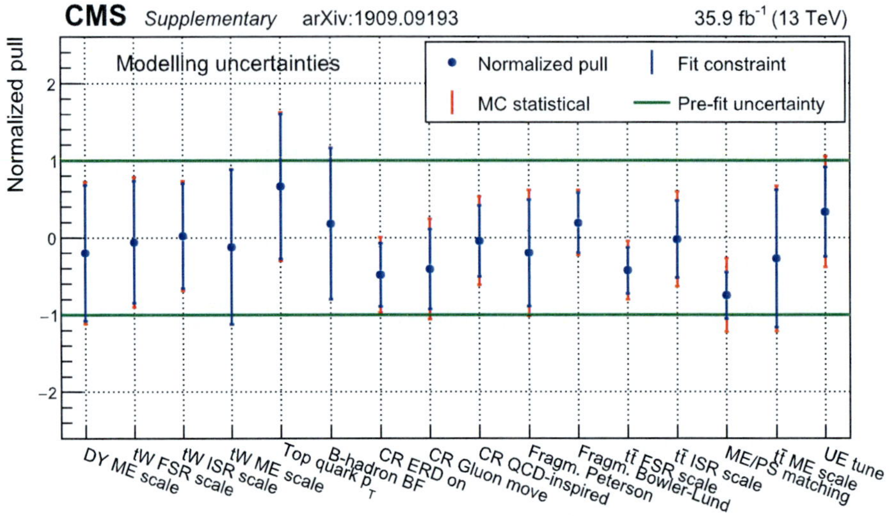

Fig. 7.4 Normalized pulls and constraints of the nuisance parameters related to the modelling uncertainties. The pulls, defined as the difference between the post-fit and pre-fit values of a nuisance parameter in units of the corresponding pre-fit uncertainty, are represented by the points, while the constraints, defined as the ratio between the post-fit and pre-fit uncertainties on a nuisance parameter, correspond to the inner vertical bars. The horizontal lines represent the pre-fit uncertainty, and the outer vertical bands indicate the additional uncertainty due to the limited statistics in the simulation, as determined using pseudo-experiments [1]

and their correlations including all experimental and modelling uncertainties are given in Table 7.3. The impact of the various sources of uncertainty on the measured $\sigma_{t\bar{t}}^{(\mu_k)}$, estimated as explained in Chap. 5, is given in Tables 7.4, 7.5, 7.6 7.7. It can be noted that, although the total uncertainties are of similar size, the impact of JES and PDF uncertainties is larger compared to the one on the inclusive cross section measured in Chap. 5. This is due to the fact that these uncertainties can introduce significant migrations between the bins of $m_{t\bar{t}}$. As expected, the uncertainty in the modelling of the top quark p_T is found to have a more significant effect at high values of $m_{t\bar{t}}$. This analysis represents the first example of an application of the maximum-likelihood unfolding method to a top-quark related measurement, and provides results with significantly improved precision with respect to those obtained by performing unfolding as a separate step, e.g. the ones of Ref. [7].

In the fit, the top quark mass in the simulation is measured to be:

$$m_t^{MC} = 172.40 \pm 0.18\,(\text{stat}) \pm^{0.67}_{0.68}\,(\text{syst})\,\text{GeV},$$

in good agreement with the result of Chap. 5 within the uncorrelated part of the uncertainty, which is estimated to be of at least 0.27 GeV.

In Fig. 7.6 the measured $\sigma_{t\bar{t}}^{(\mu_k)}$ are compared to the corresponding NLO theoretical predictions obtained assuming different values of $m_t(m_t)$. The calculations are per-

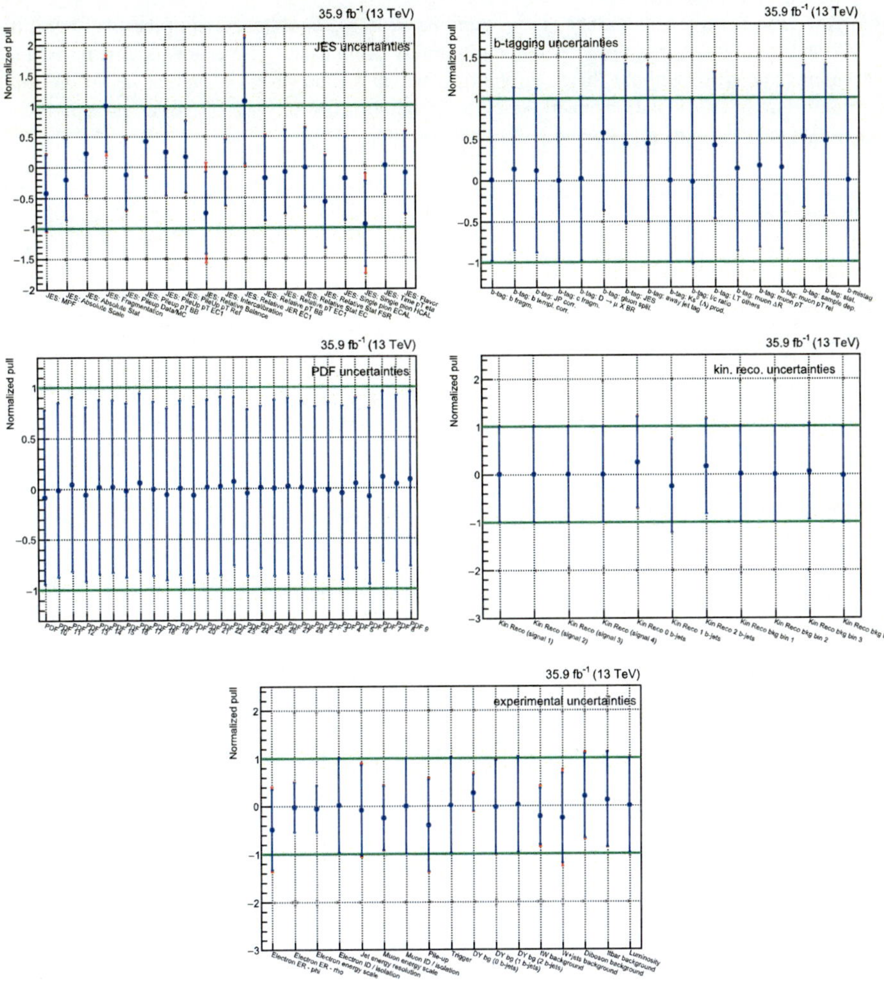

Fig. 7.5 The same as Fig. 7.4, but for the JES (upper left), b tagging (upper right), PDF (middle left), kinematic reconstruction (middle right), and other experimental uncertainties (lower)

Table 7.3 Correlations between the measured $\sigma_{t\bar{t}}^{(\mu_k)}$, including all systematic uncertainties [1]

	$\sigma_{t\bar{t}}^{(\mu_1)}$	$\sigma_{t\bar{t}}^{(\mu_2)}$	$\sigma_{t\bar{t}}^{(\mu_3)}$	$\sigma_{t\bar{t}}^{(\mu_4)}$
$\sigma_{t\bar{t}}^{(\mu_1)}$	1.00			
$\sigma_{t\bar{t}}^{(\mu_2)}$	0.64	1.00		
$\sigma_{t\bar{t}}^{(\mu_3)}$	0.72	0.60	1.00	
$\sigma_{t\bar{t}}^{(\mu_4)}$	0.32	0.65	0.47	1.00

Table 7.4 The relative uncertainty in $\sigma_{t\bar{t}}^{(\mu_1)}$ and its sources, as obtained from the likelihood fit. The MC statistical uncertainty is determined separately using pseudo-experiments. The individual uncertainties are given without their correlations, which are however accounted for in the total uncertainty. For the extrapolation uncertainties, the $\pm$ notation is used if a positive variation produces an increase in $\sigma_{t\bar{t}}^{(\mu_1)}$, while the $\mp$ notation is used otherwise [1]

Source	Uncertainty [%]
Jet energy scale	1.0
PDF	1.1
Lepton ID/isolation	2.2
Electron energy	0.5
b quark fragmentation	1.1
b tagging	0.2
Colour reconnection	0.7
Kinematic reconstruction	0.4
DY ME scale	0.4
Jet energy resolution	0.2
Muon energy scale	0.1
Pile-up	0.5
tW FSR scale	0.2
tW ISR scale	0.2
tW ME scale	0.2
m_t^{MC}	0.5
Top quark p_T	0.7
Trigger	0.3
b hadron BF	0.1
$t\bar{t}$ FSR scale	0.7
$t\bar{t}$ ISR scale	0.3
ME/PS matching	0.2
$t\bar{t}$ ME scale	0.3
UE tune	0.3
DY background	0.9
tW background	0.6
W+jets background	0.1
Diboson background	0.6
$t\bar{t}$ background	0.3
Integrated luminosity	2.6
Statistical	0.7
MC statistical	1.5
Extrapolation uncertainties	
$t\bar{t}$ ISR scale	± 0.2
$t\bar{t}$ FSR scale	± 0.1
$t\bar{t}$ ME scale	± 0.1
UE tune	$\mp^{<0.1}_{0.1}$
PDF	$\pm^{0.8}_{0.5}$
Top quark p_T	$\pm^{<0.1}_{0.1}$
Total $\sigma_{t\bar{t}}^{(\mu_1)}$ uncertainty	$^{+4.7}_{-4.4}$

Table 7.5 The same as Table 7.4, but for $\sigma_{t\bar{t}}^{(\mu_2)}$

Source	Uncertainty [%]
Jet energy scale	1.3
PDF	1.0
Lepton ID/isolation	2.3
Electron energy	0.4
b quark fragmentation	1.0
b tagging	0.5
Colour reconnection	1.0
Kinematic reconstruction	0.4
DY ME scale	0.2
Jet energy resolution	0.5
Muon energy scale	0.2
Pile-up	0.2
tW FSR scale	0.2
tW ISR scale	0.2
tW ME scale	0.2
m_t^{MC}	0.4
Top quark p_T	0.4
Trigger	0.4
b hadron BF	0.2
$t\bar{t}$ FSR scale	1.5
$t\bar{t}$ ISR scale	0.3
ME/PS matching	0.8
$t\bar{t}$ ME scale	0.8
UE tune	0.2
DY background	1.2
tW background	1.1
W+jets background	0.2
Diboson background	0.3
$t\bar{t}$ background	0.2
Integrated luminosity	2.6
Statistical	0.6
MC statistical	1.8
Extrapolation uncertainties	
$t\bar{t}$ ISR scale	$\mp^{0.2}_{0.1}$
$t\bar{t}$ FSR scale	$\mp^{<0.1}_{0.1}$
$t\bar{t}$ ME scale	$\mp^{0.1}_{0.2}$
UE tune	$\mp^{0.1}_{<0.1}$
PDF	$\pm^{0.8}_{0.6}$
Top quark p_T	± 0.1
Total $\sigma_{t\bar{t}}^{(\mu_2)}$ uncertainty	$^{+5.0}_{-4.8}$

Table 7.6 The same as Table 7.4, but for $\sigma_{t\bar{t}}^{(\mu_3)}$

Source	Uncertainty [%]
Jet energy scale	1.8
PDF	1.2
Lepton ID/isolation	2.2
Electron energy	0.5
b quark fragmentation	1.1
b tagging	0.7
Colour reconnection	0.7
Kinematic reconstruction	0.3
DY ME scale	0.1
Jet energy resolution	0.1
Muon energy scale	0.1
Pile-up	0.3
tW FSR scale	0.1
tW ISR scale	0.1
tW ME scale	0.1
m_t^{MC}	0.6
Top quark p_T	1.2
Trigger	0.4
b hadron BF	0.1
$t\bar{t}$ FSR scale	0.5
$t\bar{t}$ ISR scale	0.5
ME/PS matching	0.5
$t\bar{t}$ ME scale	0.7
UE tune	0.2
DY background	1.2
tW background	0.9
W+jets background	0.1
Diboson background	0.3
$t\bar{t}$ background	0.1
Integrated luminosity	2.6
Statistical	0.8
MC statistical	1.4
Extrapolation uncertainties	
$t\bar{t}$ ISR scale	$\pm^{0.2}_{0.1}$
$t\bar{t}$ FSR scale	$\pm^{0.2}_{<0.1}$
$t\bar{t}$ ME scale	$\mp^{0.4}_{0.5}$
UE tune	± 0.1
PDF	$\pm^{0.9}_{0.6}$
Top quark p_T	$\pm^{0.2}_{0.4}$
Total $\sigma_{t\bar{t}}^{(\mu_3)}$ uncertainty	$^{+5.0}_{-4.8}$

Table 7.7 The same as Table 7.4, but for $\sigma_{\text{t}\bar{\text{t}}}^{(\mu_4)}$

Source	Uncertainty [%]
Jet energy scale	2.2
PDF	1.5
Lepton ID/isolation	2.0
Electron energy	0.4
b quark fragmentation	0.9
b tagging	1.2
Colour reconnection	1.4
Kinematic reconstruction	0.6
DY ME scale	0.4
Jet energy resolution	0.8
Muon energy scale	0.4
Pile-up	0.4
tW FSR scale	0.4
tW ISR scale	0.4
tW ME scale	0.4
m_{t}^{MC}	0.4
Top quark p_{T}	3.5
Trigger	0.5
b hadron BF	0.4
$\text{t}\bar{\text{t}}$ FSR scale	1.3
$\text{t}\bar{\text{t}}$ ISR scale	0.4
ME/PS matching	1.0
$\text{t}\bar{\text{t}}$ ME scale	1.8
UE tune	0.8
DY background	1.6
tW background	1.0
W+jets background	0.5
Diboson background	0.4
$\text{t}\bar{\text{t}}$ background	0.4
Integrated luminosity	2.6
Statistical	1.8
MC statistical	2.5
Extrapolation uncertainties	
$\text{t}\bar{\text{t}}$ ISR scale	$\mp^{0.8}_{0.7}$
$\text{t}\bar{\text{t}}$ FSR scale	$\pm^{0.2}_{<0.1}$
$\text{t}\bar{\text{t}}$ ME scale	$\mp^{0.8}_{1.2}$
UE tune	$\pm^{0.1}_{0.2}$
PDF	$\pm^{1.2}_{0.9}$
Top quark p_{T}	$\pm^{0.6}_{1.2}$
Total $\sigma_{\text{t}\bar{\text{t}}}^{(\mu_4)}$ uncertainty	$^{+7.2}_{-6.9}$

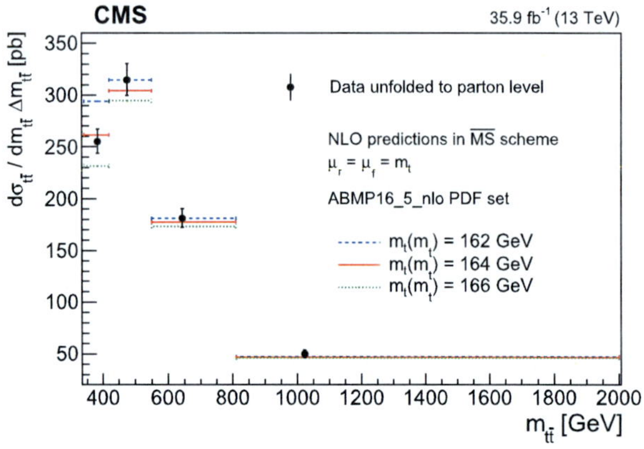

Fig. 7.6 Measured values of $\sigma_{t\bar{t}}^{(\mu_k)}$ (points) and their uncertainties (vertical error bars) compared to NLO predictions in the $\overline{\text{MS}}$ scheme obtained with different values of $m_t(m_t)$ (horizontal lines of different styles). The values of $\sigma_{t\bar{t}}^{(\mu_k)}$ are shown at the representative scale of the process μ_k, defined as the centre-of-gravity of bin k in $m_{t\bar{t}}$. The first and last bins contain all events with $m_{t\bar{t}} < 420\,\text{GeV}$ and $m_{t\bar{t}} > 810\,\text{GeV}$, respectively [1]

formed using a version [5] of the MCFM v6.8 program [9, 10] where the heavy quark masses are treated in the $\overline{\text{MS}}$ scheme. The calculation is interfaced to the ABMP16 NLO PDF set [11], in which the correlation between α_S and the gluon PDF is taken into account. In the calculation, the value of $\alpha_S(m_Z)$ is set to 0.1191, which corresponds to the value determined at NLO in the PDF fit [11]. The renormalization and factorization scales are both set to m_t, and five active flavours ($n_f = 5$) are assumed both in the calculation and in the PDFs. It can be noted from Fig. 7.6 that the sensitivity to the top quark mass is higher at low values of $m_{t\bar{t}}$, while it decreases at higher $m_{t\bar{t}}$ where m_t becomes negligible with respect to the scale of the process.

The measurement presented in this section is cross-checked by performing the analysis using an alternative kinematic reconstruction algorithm that does not depend on the parameter m_t^{kin}. The results, presented in Appendix E, are found to be in agreement with the ones presented in this chapter, although with slightly larger uncertainties.

7.3 Extraction of the Running

The measured $\sigma_{t\bar{t}}^{(\mu_k)}$ and the theoretical predictions described in the previous section are used to extract the running of $m_t(\mu)$ as a function of $\mu = m_{t\bar{t}}$. In each bin of $m_{t\bar{t}}$ independently, the value of $m_t(m_t)$ is extracted by means of a χ^2 fit of the theoretical predictions to the measured cross section. A signed χ^2 that takes into account any asymmetry in the input uncertainties is used, defined following Ref. [12]:

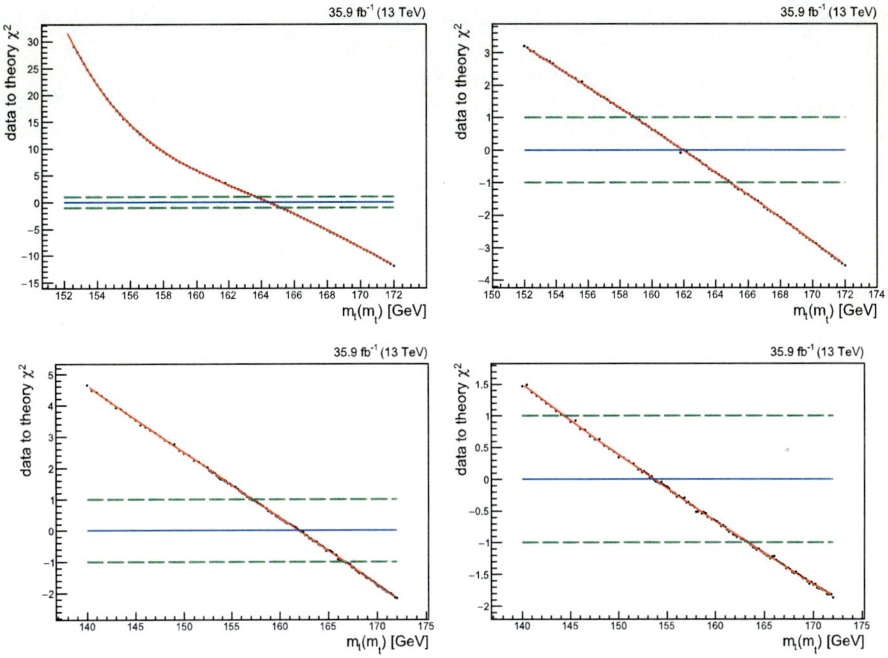

Fig. 7.7 Values of $\sqrt{\chi^2_k}$ as a function of $m_t(m_t)$ for the first (upper left), second (upper right), third (lower left) and fourth (lower right) bin in $m_{t\bar{t}}$. Only the fit uncertainties are included in the definition of the χ^2. The horizontal lines represent the conditions $\sqrt{\chi^2_k} = 0$ and $\sqrt{\chi^2_k} = \pm 1$, which are used to determine the extracted values of $m_t(m_t)$ and their uncertainties, respectively

$$\sqrt{\chi^2_k(m_t)} = \frac{\delta_k}{\Delta\sigma_k}\sqrt{1 - 2A_k\frac{\delta_k}{\Delta\sigma_k} + 5A_k^2\left(\frac{\delta_k}{\Delta\sigma_k}\right)^2}. \tag{7.4}$$

Here, $\delta_k = \delta_k(m_t)$ is the signed difference between the predicted and measured $\sigma^{(\mu_k)}_{t\bar{t}}$, $\Delta\sigma_k = (\Delta\sigma_k^+ + \Delta\sigma_k^-)/2$ is the average between the upper ($\Delta\sigma_k^+$) and lower ($\Delta\sigma_k^-$) uncertainty in $\sigma^{(\mu_k)}_{t\bar{t}}$, and $A_k = (\Delta\sigma_k^+ - \Delta\sigma_k^-)/(\Delta\sigma_k^+ + \Delta\sigma_k^-)$ is the relative asymmetry in the uncertainty. With this definition, the sign of $\sqrt{\chi^2_k}$ corresponds to the sign of δ_k. The extracted values of $m_t(m_t)$ and their uncertainties are then evolved to the corresponding scale μ_k using RGE solutions at one-loop precision with $n_f = 5$, using the CRunDec v3.0 program [13]. In the conversion, the value of α_S is set to the one used in the calculation. The values of $\sqrt{\chi^2_k}$ in the different bins of $m_{t\bar{t}}$ obtained as a function of $m_t(m_t)$ are shown in Fig. 7.7, where the dependence is parametrized using a fourth-order polynomial function. The best fit values of $m_t(m_t)$ correspond to the values for which $\sqrt{\chi^2_k} = 0$, while their uncertainties are determined by the condition $\sqrt{\chi^2_k} = \pm 1$. The extracted values of $m_t(m_t)$ and their conversion to the

Table 7.8 Extracted values of $m_t(m_t)$ in the different bins of $m_{t\bar{t}}$ and their conversion to $m_t(\mu_k)$ obtained using RGE solutions at one-loop precision with five active flavours. The reported uncertainties in $m_t(m_t)$ and $m_t(\mu_k)$ include the fit uncertainties only

μ_k [GeV]	$m_t(m_t)$ [GeV]	$m_t(\mu_k)$ [GeV]
384	$164.41\,^{+0.75}_{-0.77}$	$155.44\,^{+0.75}_{-0.78}$
476	161.9 ± 3.0	$150.9\,^{+2.9}_{-3.0}$
644	$162.0\,^{+4.7}_{-4.8}$	$148.2\,^{+4.6}_{-4.7}$
1024	$153.6\,^{+9.7}_{-9.3}$	$136.4\,^{+9.2}_{-8.8}$

corresponding $m_t(\mu_k)$ are summarized in Table 7.8. In the table, the uncertainty only includes the experimental and modelling uncertainties in $\sigma_{t\bar{t}}^{(\mu_k)}$ in the visible phase space, which are referred to as fit uncertainties. The values of $m_t(m_t)$ extracted in the different bins of $m_{t\bar{t}}$ are found to be in good agreement with each other. Since the fit uncertainties are found to be largely symmetric, their values are symmetrized as the average of the up and down uncertainties.

The impact of PDF and extrapolation uncertainties on the extracted masses is estimated using the PDF eigenvectors of ABMP16 and the results of Sect. 7.2, respectively. The up and down variations are separately summed in quadrature to yield the total uncertainties. As noted in Chap. 6, the eigenvectors of the ABMP16 PDF set incorporate the uncertainty in α_S. The impact of the α_S uncertainty on the CRunDec conversion is estimated by varying the value of $\alpha_S(m_Z)$ extracted in the ABMP16 fit within its uncertainty [11], and was found to be negligible. In Fig. 7.8, the extracted values $m_t(\mu_k)$ are compared to the value of $m_t(m_t)$ obtained from the inclusive cross section measured in Chap. 5 using NLO Hathor predictions and the ABMP16 PDF set, which is indicated with $m_t^{incl}(m_t)$. The scale uncertainties are estimated by varying the renormalization and factorization scales in the calculation by a factor of two, avoiding cases for which $\mu_r/\mu_f = 4$ or $1/4$. In Fig. 7.8, good agreement between the extracted values of $m_t(\mu_k)$ and the evolved uncertainty band of $m_t^{incl}(m_t)$ is observed. However, it should be noted that the value of $m_t^{incl}(m_t)$ is determined using the same data set. As expected, the sensitivity on m_t decreases with increasing $m_{t\bar{t}}$, resulting in higher uncertainty in the extracted mass. The numerical values of $m_t(\mu_k)$ are determined to be:

$$
\begin{aligned}
m_t(\mu_1) &= 155.4 \pm 0.8\ (\text{fit}) \pm 0.2\ (\text{PDF} + \alpha_S) \pm 0.1\ (\text{extr})\ ^{+0.9}_{-0.6}\ (\text{scale}),\\
m_t(\mu_2) &= 150.9 \pm 3.0\ (\text{fit})\ ^{+1.1}_{-0.7}\ (\text{PDF} + \alpha_S)\ ^{+0.4}_{-0.5}\ (\text{extr})\ ^{+3.9}_{-4.3}\ (\text{scale}),\\
m_t(\mu_3) &= 148.2 \pm 4.6\ (\text{fit})\ ^{+2.0}_{-1.4}\ (\text{PDF} + \alpha_S)\ ^{+0.9}_{-1.0}\ (\text{extr})\ ^{+7.3}_{-9.5}\ (\text{scale}),\\
m_t(\mu_4) &= 136.4 \pm 9.0\ (\text{fit})\ ^{+3.8}_{-3.0}\ (\text{PDF} + \alpha_S)\ ^{+2.8}_{-2.3}\ (\text{extr})\ ^{+9.6}_{-16.1}\ (\text{scale}),
\end{aligned}
$$

while the value of $m_t^{incl}(m_t)$ is found to be:

$$
m_t^{incl}(m_t) = 162.9 \pm 1.6\ (\text{fit} + \text{extr} + \text{PDF} + \alpha_S)\ ^{+2.5}_{-3.0}\ (\text{scale}).
$$

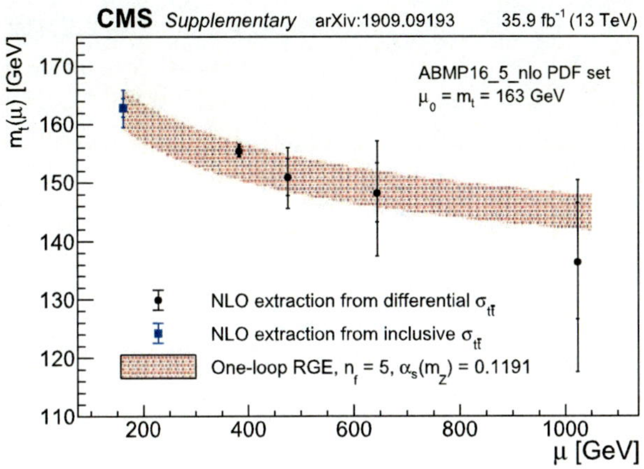

Fig. 7.8 Values of $m_{\mathrm{t}}(\mu_k)$ extracted from the measured $\mathrm{d}\sigma_{\mathrm{t\bar{t}}}/\mathrm{d}m_{\mathrm{t\bar{t}}}$ (points), compared to the value of $m_{\mathrm{t}}(m_{\mathrm{t}})$ extracted from the inclusive $\sigma_{\mathrm{t\bar{t}}}$ (square). The inner vertical bands correspond to the combination of fit, extrapolation, and PDF uncertainties, while the outer vertical bands include the contribution of the scale uncertainties [1]. In this plot, the RGE evolution is calculated from the initial scale $\mu_0 = m_{\mathrm{t}}^{\mathrm{incl}}(m_{\mathrm{t}}) = 163\,\mathrm{GeV}$

In order to benefit from the cancellation of correlated uncertainties in the extracted $m_{\mathrm{t}}(\mu_k)$, the running is determined with respect to a reference scale μ_{ref}. The slope of the scale dependence of m_{t}, defined as $r(\mu) = m_{\mathrm{t}}(\mu)/m_{\mathrm{t}}(\mu_{\mathrm{ref}})$, is interpreted as the running of the top quark mass. The reference scale is arbitrarily chosen to be $\mu_{\mathrm{ref}} = \mu_2$, and the quantities $r_{k2} = m_{\mathrm{t}}(\mu_k)/m_{\mathrm{t}}(\mu_2)$ are experimentally determined. This choice of μ_{ref} was found to minimize the correlations between the extracted ratios. However, the result of the measurement does not depend on the choice of the reference scale. Another advantage of this approach is that the quantity $r(\mu)$, unlike $m_{\mathrm{t}}(\mu)$, solely depends on the solution of the RGE, and not on the absolute value of the top quark mass. Furthermore, scale uncertainties are not considered in the determination of r_{k2}, since the running is investigated at a fixed order in perturbation theory. In fact, scale variations are normally used to estimate the effect of higher order corrections, and therefore not applicable in this case.

In order to determine the ratios r_{k2}, the correlations between the values of $m_{\mathrm{t}}(\mu_k)$ have to be estimated and taken into account. The correlations introduced by the extrapolation and PDF uncertainties are automatically accounted for by performing simultaneous variations of the $\sigma_{\mathrm{t\bar{t}}}^{(\mu_k)}$ and PDFs. The resulting up and down uncertainties in each r_{k2} are then summed in quadrature to yield the total extrapolation and PDF uncertainties. The correlations introduced by the fit uncertainties, instead, are estimated by generating pseudo-experiments of $\sigma_{\mathrm{t\bar{t}}}^{(\mu_k)}$, according to the full covariance matrix. With this procedure it is verified that the correlations between the extracted $m_{\mathrm{t}}(\mu_k)$ correspond to those in the measured $\sigma_{\mathrm{t\bar{t}}}^{(\mu_k)}$ (Table 7.3).

The extracted ratios are compared to the corresponding RGE solution for $r(\mu)$ in Fig. 7.9, and to the value of $m_{\mathrm{t}}^{\mathrm{incl}}(m_{\mathrm{t}})/m_{\mathrm{t}}(\mu_2)$ in Fig. 7.10. Here, the uncertainty in $m_{\mathrm{t}}^{\mathrm{incl}}(m_{\mathrm{t}})$ includes fit, extrapolation and PDF uncertainties, while the value of

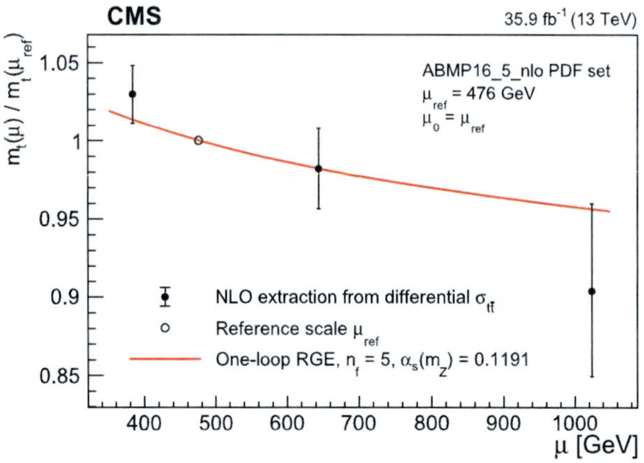

Fig. 7.9 Extracted running of the top quark mass $m_t(\mu_k)/m_t(\mu_{ref})$ (filled points) compared to the RGE prediction at one-loop precision with $n_f = 5$ (line), evolved from the initial scale $\mu_0 = \mu_{ref} = 476$ GeV (open point) [1]

$m_t(\mu_2)$ in the ratio $m_t^{incl}(m_t)/m_t(\mu_2)$ is taken without uncertainty. In both cases, good agreement with the RGE prediction is observed over the investigated range. The values of the ratios are determined to be:

$$
\begin{aligned}
r_{12} &= 1.030 \pm 0.018 \text{ (fit) } {}^{+0.003}_{-0.006} \text{ (PDF} + \alpha_S) {}^{+0.003}_{-0.002} \text{ (extr)}, \\
r_{32} &= 0.982 \pm 0.025 \text{ (fit) } {}^{+0.006}_{-0.005} \text{ (PDF} + \alpha_S) \pm 0.004 \text{ (extr)}, \\
r_{42} &= 0.904 \pm 0.050 \text{ (fit) } {}^{+0.019}_{-0.017} \text{ (PDF} + \alpha_S) {}^{+0.017}_{-0.013} \text{ (extr)}.
\end{aligned}
$$

The impact of the most significant sources of systematic uncertainty on the measured ratios is summarized in Table 7.9. These quantities are estimated by varying the corresponding nuisance parameters in the fit by their post-fit uncertainty, taking the correlation between the $\sigma_{t\bar{t}}^{(\mu_k)}$ into account. As expected, the JES and modelling uncertainty play a significant role, since they can introduce migrations between the $m_{t\bar{t}}$ bins. Systematic uncertainties related to the normalization of the processes, such as the integrated luminosity and the lepton identification and isolation scale factors, are also found to play a significant role. This is explained considering that the uncertainties in the different $m_t(\mu_k)$ vary significantly as a function of $m_{t\bar{t}}$, and only part of the uncertainty cancels in the ratios. The impact of the extrapolation uncertainties on the extracted ratios is shown in Table 7.10.

The correlations between the extracted ratios introduced by the fit uncertainties are estimated with the same pseudo-experiment procedure used for the $m_t(\mu_k)$. The obtained correlation plots are shown in Fig. 7.11. Having defined ρ_{ik} as the correlation between r_{i2} and r_{k2}, the correlations are found to be:

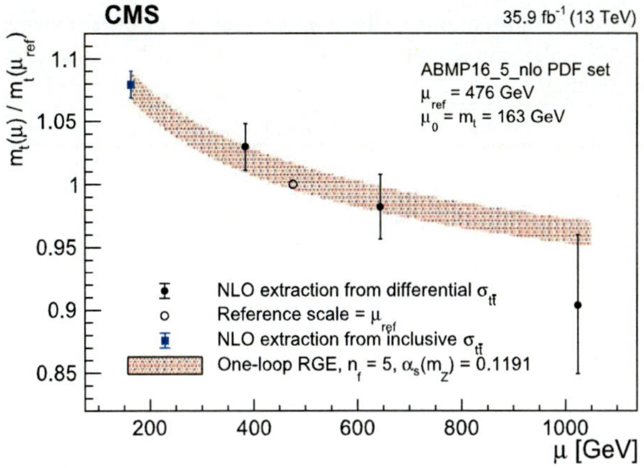

Fig. 7.10 The result of Fig. 7.9 is compared to the value of $m_{\mathrm{t}}^{\mathrm{incl}}(m_{\mathrm{t}})/m_{\mathrm{t}}(\mu_{\mathrm{ref}})$ (square), where $m_{\mathrm{t}}^{\mathrm{incl}}(m_{\mathrm{t}})$ is the value of $m_{\mathrm{t}}(m_{\mathrm{t}})$ extracted from the inclusive cross section measured in Chap. 5, which is based on the same data set. The uncertainty in $m_{\mathrm{t}}^{\mathrm{incl}}(m_{\mathrm{t}})$ is evolved from the initial scale $\mu_0 = m_{\mathrm{t}} = 163\,\mathrm{GeV}$ using the RGE prediction at one-loop precision with $n_f = 5$ [1]

$$\rho_{13} = 0.13,$$
$$\rho_{34} = 0.11,$$
$$\rho_{14} = -0.45.$$

Finally, the agreement of the result with the RGE prediction and the significance of the observed running are quantified by parametrizing the extracted ratios with the function:

$$f(x, \mu) = x\,[r(\mu) - 1] + 1, \tag{7.5}$$

which interpolates between the RGE prediction $r(\mu)$ and a hypothetical no-running scenario, i.e. $r(\mu) = 1$. In fact, the function in Eq. 7.5 corresponds to $r(\mu)$ for $x = 1$, and to 1 for $x = 0$. The best-fit value of x, indicated with $\hat{x}$, is determined via a χ^2 fit of the function $f(x, \mu)$ to the measured r_{k2}, taking all correlations into account. As in the determination of $m_{\mathrm{t}}(\mu_k)$ and r_{k2}, the χ^2 definition only includes the fit uncertainties, while extrapolation and PDF uncertainties are estimated separately. The χ^2 profile as a function of x is shown in Fig. 7.12, and the best-fit value of x is determined to be:

$$\hat{x} = 2.05 \pm 0.61\ (\mathrm{fit})\ {}^{+0.31}_{-0.55}\ (\mathrm{PDF} + \alpha_S)\ {}^{+0.24}_{-0.49}\ (\mathrm{extr}).$$

Based on this result, the observed running is found to be in agreement with the one-loop RGE prediction within 1.1 standard deviation, and excludes the no-running hypothesis at above 95% confidence level.

Table 7.9 Most relevant contributions to the total uncertainty in the extracted r_{k2}. The sources are grouped in energy scales and resolutions, overall normalization, modelling, and background normalization uncertainties

Source	$\Delta r_{12}[\%]$	$\Delta r_{32}[\%]$	$\Delta r_{42}[\%]$
Jet energy resolution	0.19	0.10	0.59
JES Fragmentation	0.08	0.44	0.09
JES Relative JER EC1	0.38	0.71	1.42
JES Relative p_T BB	0.08	0.20	0.78
JES Relative p_T EC1	0.09	0.33	0.02
JES Relative Stat FSR	0.07	0.24	0.79
JES Single pion HCAL	0.16	0.32	0.66
Electron energy resolution	0.20	0.43	0.26
Electron ID/isolation	0.56	0.44	1.02
Muon ID/isolation	0.39	0.24	0.31
Integrated luminosity	0.78	0.64	1.47
CR ERD on	0.39	0.01	0.52
CR Gluon move	0.02	0.20	1.10
CR QCD-inspired	0.27	0.03	0.05
m_t^{MC}	0.20	0.56	0.36
Top quark p_T	0.22	0.65	3.25
b quark fragmentation	0.30	0.26	0.22
Peterson fragmentation	0.11	0.06	0.81
$t\bar{t}$ FSR scale	0.54	0.29	0.63
$t\bar{t}$ ISR scale	0.07	0.42	0.08
ME/PS matching	0.34	0.64	0.71
$t\bar{t}$ ME scale	0.33	0.15	1.42
UE tune	0.09	0.07	0.74
PDF 22	0.07	0.17	0.66
tW background	0.40	0.12	0.54
DY background (0 b jets)	0.41	0.26	1.04

Table 7.10 Impact of extrapolation uncertainties on the extracted ratios. The $\pm$ notation is used if an up variation of the corresponding nuisance parameter results in an increase in the extracted ratio, while the $\mp$ notation is used otherwise

Source	$\Delta r_{12}[\%]$	$\Delta r_{32}[\%]$	$\Delta r_{42}[\%]$
$t\bar{t}$ ISR scale	$\mp^{0.06}_{0.11}$	∓ 0.15	± 0.7
$t\bar{t}$ FSR scale	$\mp^{0.05}_{0.01}$	$\mp^{0.16}_{<0.01}$	$\mp^{0.20}_{0.03}$
$t\bar{t}$ ME scale	$\mp^{0.07}_{0.05}$	$\pm^{0.26}_{0.19}$	$\pm^{1.11}_{0.71}$
UE Tune	$\pm^{0.01}_{0.03}$	$\mp^{0.05}_{0.09}$	$\mp^{0.10}_{0.23}$
PDF	$\pm^{0.25}_{0.17}$	$\mp^{0.26}_{0.19}$	$\mp^{0.82}_{0.63}$
Top quark p_T	$\pm^{0.03}_{0.05}$	$\mp^{0.11}_{0.21}$	$\mp^{0.59}_{1.15}$

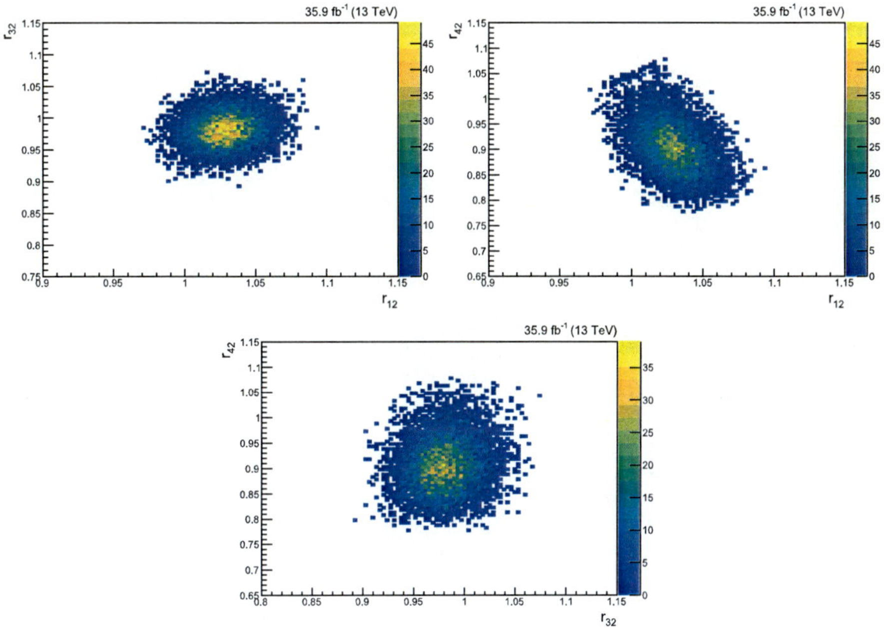

Fig. 7.11 Plots used to estimate the correlations between r_{12} and r_{32} (upper left), r_{12} and r_{42} (upper right), and r_{32} and r_{42} (lower), obtained using pseudo-experiments

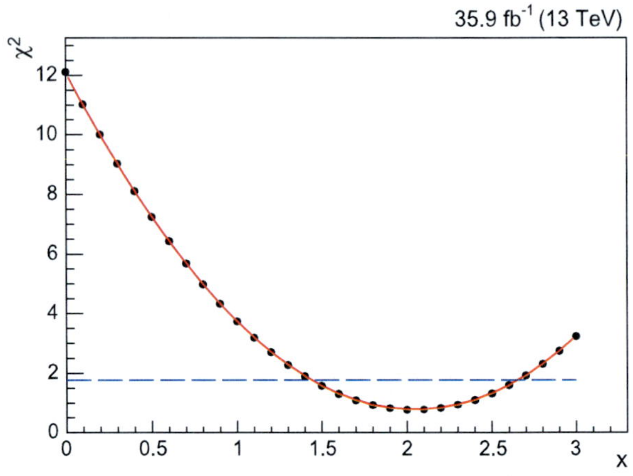

Fig. 7.12 The χ^2 profile as a function of the parameter x defined in Eq. 7.5, parametrized with a second-order polynomial function. The horizontal line represents the minimum value of the $\chi^2 + 1$, which defines the fit uncertainty

References

1. *CMS* (2020) Collaboration, Running of the top quark mass from proton-proton collisions at $\sqrt{s} = 13\,\text{TeV}$. Phys Lett B **803**, 135263 https://doi.org/10.1016/j.physletb.2020.135263, arXiv:1909.09193
2. *CMS* (2017) Collaboration, Measurement and QCD analysis of double-differential inclusive jet cross sections in pp collisions at $\sqrt{s} = 8\,\text{TeV}$ and cross section ratios to 2.76 and 7TeV. JHEP 03:156. https://doi.org/10.1007/JHEP03(2017)156, arXiv:1609.05331
3. Bethke S (2004) QCD studies at LEP. Phys Rept **403**:203. https://doi.org/10.1016/j.physrep.2004.08.014. arXiv:hep-ex/0406058
4. Gizhko A et al (2017) Running of the charm-quark mass from HERA deep-inelastic scattering data. Phys Lett B 775:233. https://doi.org/10.1016/j.physletb.2017.11.002, arXiv:1705.08863
5. Dowling M, Moch S (2014) Differential distributions for top-quark hadro-production with a running mass. Eur Phys J C 74:3167 .https://doi.org/10.1140/epjc/s10052-014-3167-x, arXiv:1305.6422
6. *CMS* (2015) Collaboration, Measurement of the differential cross section for top quark pair production in pp collisions at $\sqrt{s} = 8\,\text{TeV}$. Eur Phys J C 75:542. https://doi.org/10.1140/epjc/s10052-015-3709-x, arXiv:1505.04480
7. *CMS* (2019) Collaboration, Measurements of $\text{t}\bar{\text{t}}$ differential cross sections in proton-proton collisions at $\sqrt{s} = 13\text{TeV}$ using events containing two leptons. JHEP 02:149. https://doi.org/10.1007/JHEP02(2019)149, arXiv:1811.06625
8. *CDF* (2006) Collaboration, Measurement of the top quark mass using template methods on dilepton events in proton antiproton collisions at $\sqrt{s} = 1.96\,\text{TeV}$. Phys Rev D 73:112006. https://doi.org/10.1103/PhysRevD.73.112006, arXiv:hep-ex/0602008
9. Campbell JM, Ellis RK (2010) MCFM for the Tevatron and the LHC. Nucl Phys B Proc Suppl 205–206:10 https://doi.org/10.1016/j.nuclphysbps.2010.08.011, arXiv:1007.3492
10. Campbell JM, Ellis RK (2015) Top-quark processes at NLO in production and decay. J Phys G 42:015005. https://doi.org/10.1088/0954-3899/42/1/015005, arXiv:1204.1513
11. Alekhin S, Blümlein J, Moch S (2018) NLO PDFs from the ABMP16 fit. Eur Phys J C78:477. https://doi.org/10.1140/epjc/s10052-018-5947-1, arXiv:1803.07537
12. Barlow R (2003) Asymmetric errors. In: Statistical problems in particle physics, astrophysics and cosmology. Proceedings, Conference, PHYSTAT 2003, Stanford, USA, p WEMT002, Sept 2003. arXiv:physics/0401042
13. Schmidt B, Steinhauser, M (2012) CRunDec: a C++ package for running and decoupling of the strong coupling and quark masses. Comput Phys Commun 183:1845 (2012). https://doi.org/10.1016/j.cpc.2012.03.023, arXiv:1201.6149

Chapter 8
Summary and Conclusions

Several new results related to the $\mathrm{t\bar{t}}$ production cross section and the top quark mass are presented in this thesis. The measurements are performed using proton-proton collision data at the centre-of-mass energy of 13 TeV collected by the CMS detector at the CERN LHC in 2016, corresponding to an integrated luminosity of 35.9 fb^{-1}. In this work, the inclusive $\mathrm{t\bar{t}}$ production cross section is measured simultaneously with $m_{\mathrm{t}}^{\mathrm{MC}}$, and the result is used to determine the value of $m_{\mathrm{t}}(m_{\mathrm{t}})$, $m_{\mathrm{t}}^{\mathrm{pole}}$, and $\alpha_{\mathrm{S}}(m_{\mathrm{Z}})$ at NNLO. A similar method is then used to perform the first measurement of the running of the top quark mass as defined in the $\overline{\mathrm{MS}}$ renormalization scheme.

The simultaneous measurement of $\sigma_{\mathrm{t\bar{t}}}$ and $m_{\mathrm{t}}^{\mathrm{MC}}$ is performed by means of a maximum-likelihood fit to multidifferential final-state distributions. The systematic uncertainties are treated as nuisance parameters of the fit, and are constrained within the visible phase space. Candidate $\mathrm{t\bar{t}}$ events are selected in the final state with an electron and a muon of opposite charge and are categorized based on the multiplicity of jets and b-tagged jets. The dependence of the measured $\sigma_{\mathrm{t\bar{t}}}$ on the value of $m_{\mathrm{t}}^{\mathrm{MC}}$ is mitigated in the fit, where $m_{\mathrm{t}}^{\mathrm{MC}}$ is treated as a free parameter. The value of $m_{\mathrm{t}}^{\mathrm{MC}}$ is constrained using the $m_{\ell\mathrm{b}}^{\mathrm{min}}$ distribution, i.e. the minimum invariant mass found when combining a lepton with a b-tagged jets. A fully-consistent method to estimate the impact of the statistical uncertainty in the simulation has been developed for the purpose of this analysis. The effect is found to be large in the case of $m_{\mathrm{t}}^{\mathrm{MC}}$, which is determined from the shape of final-state distributions, while it is less severe in the case of $\sigma_{\mathrm{t\bar{t}}}$, which mainly depends on the normalization of the $\mathrm{t\bar{t}}$ process. The values of $\sigma_{\mathrm{t\bar{t}}}$ and $m_{\mathrm{t}}^{\mathrm{MC}}$ are measured to be:

$$\sigma_{\mathrm{t\bar{t}}} = 815 \pm 2\,(\mathrm{stat}) \pm 29\,(\mathrm{syst}) \pm 20\,(\mathrm{lum})\,\mathrm{pb},$$
$$m_{\mathrm{t}}^{\mathrm{MC}} = 172.33 \pm 0.14\,(\mathrm{stat})\, \pm^{0.66}_{0.72}\,(\mathrm{syst})\,\mathrm{GeV}.$$

The main sources of uncertainty in the measured $\sigma_{\mathrm{t\bar{t}}}$ are related to the integrated luminosity and the lepton identification efficiencies, while the measurement of $m_{\mathrm{t}}^{\mathrm{MC}}$

M. M. Defranchis, *First Measurement of the Running of the Top Quark Mass*, Springer Theses, https://doi.org/10.1007/978-3-030-90376-3_8

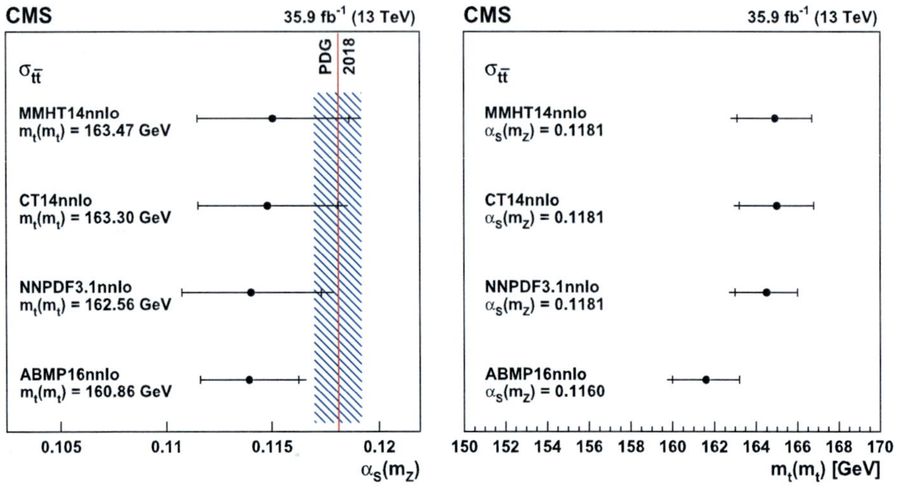

Fig. 8.1 Values of $\alpha_S(m_Z)$ (left) and $m_t(m_t)$ (right) determined via a χ^2 fit of NNLO theoretical predictions in the $\overline{\text{MS}}$ scheme to the measured $\sigma_{t\bar{t}}$, using different PDFs [1]

is limited by the jet energy scale and MC statistical uncertainties. The measured value of m_t^{MC} represents the most precise result obtained the dileptonic final state, to date, and reaches a similar level of precision as the ones obtained in the semileptonic and fully-hadronic channels.

The measured $\sigma_{t\bar{t}}$ is then used to extract $\alpha_S(m_Z)$ and $m_t(m_t)$ by comparing to NNLO theoretical predictions in the $\overline{\text{MS}}$ scheme obtained with different sets of PDFs. The results are shown in Fig. 8.1 and represent, respectively, the most precise determination of $\alpha_S(m_Z)$ at a hadron collider and the most precise determination of $m_t(m_t)$, to date. The same method is also used to extract the value of m_t^{pole} at NNLO+NNLL precision. However, due to the slower perturbative convergence of calculations of $\sigma_{t\bar{t}}$ in the on-shell scheme, larger uncertainties compared to $m_t(m_t)$ are obtained, despite the inclusion of NNLL corrections. The results for $\sigma_{t\bar{t}}$, m_t^{MC}, $\alpha_S(m_Z)$, $m_t(m_t)$, and m_t^{pole} are also documented in Ref. [1].

The method used to measure $\sigma_{t\bar{t}}$ is then extended to determine the differential $t\bar{t}$ cross section as a function of the invariant mass of the $t\bar{t}$ system, $m_{t\bar{t}}$. The differential cross section $d\sigma_{t\bar{t}}/dm_{t\bar{t}}$ is determined directly at the parton level by embedding the response matrix in the likelihood function. In the analysis, the invariant mass of the $t\bar{t}$ system is reconstructed using an analytical kinematic reconstruction method. The fit is performed in categories of b-tagged jet multiplicity and bins of reconstructed $m_{t\bar{t}}$. In Fig. 8.2 (left) the measured $d\sigma_{t\bar{t}}/dm_{t\bar{t}}$ is compared to NLO theoretical predictions in the $\overline{\text{MS}}$ scheme obtained with different values of $m_t(m_t)$.

The measured $d\sigma_{t\bar{t}}/dm_{t\bar{t}}$ and the corresponding theoretical predictions are then used to determine the value of $m_t(m_t)$ in each bin of $m_{t\bar{t}}$, independently. The extracted values of $m_t(m_t)$ are converted to the corresponding $m_t(\mu_k)$ using one-loop solutions of the RGE, where μ_k is the representative energy scale of bin k in $m_{t\bar{t}}$. In this analysis,

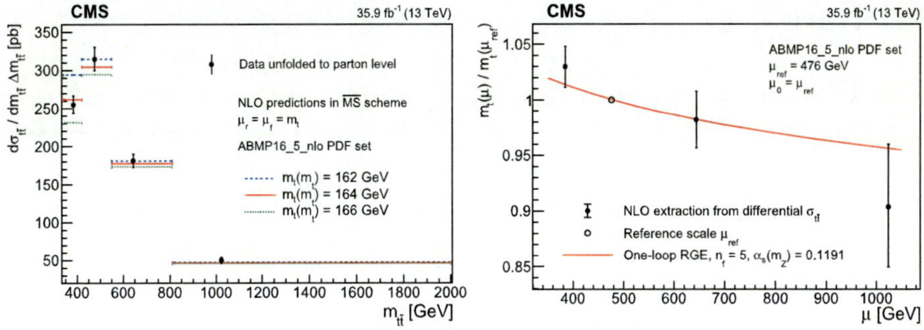

Fig. 8.2 Left: measured $d\sigma_{t\bar{t}}/dm_{t\bar{t}}$ compared to NLO theoretical predictions in the $\overline{\text{MS}}$ scheme obtained with different values of $m_t(m_t)$. Right: extracted running of the top quark mass compared to one-loop solution of the corresponding RGE [2]

μ_k is defined as the centre-of-gravity of bin k. However, the final result does not depend on the exact choice of μ_k, provided that it represents the physical energy scale of the process in that bin.

In order to benefit from the cancellation of correlated systematic uncertainties in the extracted $m_t(\mu_k)$, the running is determined with respect to a reference scale μ_{ref}, which is arbitrarily chosen to be $\mu_{\text{ref}} = \mu_2 = 476\,\text{GeV}$. The quantities $r_{k2} = m_t(\mu_k)/m_t(\mu_2)$ are experimentally determined, and the results are compared to the RGE prediction for $r(\mu) = m_t(\mu)/m_t(\mu_{\text{ref}})$. Another advantage of this method is that the quantity $r(\mu)$, unlike $m_t(\mu)$, solely depends on the considered RGE. The result is shown in Fig. 8.2 (right). The measured running is found to be compatible with the one-loop RGE solution within 1.1 standard deviations, and a hypothetical no-running scenario, i.e. $r(\mu) = 1$, is excluded at above 95% confidence level. This analysis, entirely developed and performed for the purpose of this thesis, is also documented in Ref. [2]. The precision of the measurement is limited by the uncertainties in the jet energy corrections, the integrated luminosity, the lepton identification efficiency, and the modelling of the $t\bar{t}$ signal. The result could be further improved by using theoretical calculations beyond NLO, which are currently not available in the $\overline{\text{MS}}$ scheme.

In this thesis, the running of the top quark mass is investigated for the first time, and is probed up to a scale of the order of 1 TeV. This result represents an important proof of the validity of perturbative QCD, and complements the studies on running of the bottom and charm quark masses performed at previous experiments. These results can also be used to probe the validity of theories beyond the SM that imply modifications of the RGE solutions, such as supersymmetric theories [3] or scenarios where the fermion masses are generated dynamically [4].

References

1. *CMS* Collaboration (2019) Measurement of the $t\bar{t}$ production cross section, the top quark mass, and the strong coupling constant using dilepton events in pp collisions at $\sqrt{s} = 13\text{TeV}$. Eur Phys J C 79:368. https://doi.org/10.1140/epjc/s10052-019-6863-8. arXiv:1812.10505
2. *CMS* Collaboration (2020) Running of the top quark mass from proton-proton collisions at $\sqrt{s} = 13\text{TeV}$. Phys Lett B 803:135263. https://doi.org/10.1016/j.physletb.2020.135263. arXiv:1909.09193
3. Mihaila L (2013) Precision calculations in supersymmetric theories. Adv High Energy Phys 2013:607807. https://doi.org/10.1155/2013/607807. arXiv:1310.6178
4. Christensen ND, Shrock R (2005) Implications of dynamical generation of standard-model fermion masses. Phys Rev Lett 94:241801. https://doi.org/10.1103/PhysRevLett.94.241801. arXiv:hep-ph/0501294

Appendix A
Impact of the Modelling Uncertainties on the Fit Distributions

In this appendix, the impact of the modelling uncertainties on the distributions used as inputs to the fit of Chap. 5, before and after the fit, are shown. Each figure contains three sets of plots: the upper one corresponds to the category with zero or more than two b-tagged jets, the middle one to the category with one b-tagged jets, and the lower one to the category with two b-tagged jets. For each category of b-tagged jet multiplicity, subcategories of jet multiplicity are formed, as indicated in the figures.

In the pre-fit plots, the black lines corresponds to the nominal template, while the red and blue ones represent, respectively, the templates derived by an up and down variation of the modelling uncertainty indicated in the caption. For one-sided variations, the template corresponding to the down variation is identical to the nominal one. The vertical bars represent the statistical uncertainty of the up and down templates, while the grey band represents the statistical uncertainty of the nominal template. In the post-fit distributions, the black markers correspond to the fitted template, the red and blue templates correspond to the original up and down variations, respectively, and the grey band represents the impact of the source indicated in the caption after the fit.

M. M. Defranchis, *First Measurement of the Running of the Top Quark Mass*, Springer Theses, https://doi.org/10.1007/978-3-030-90376-3

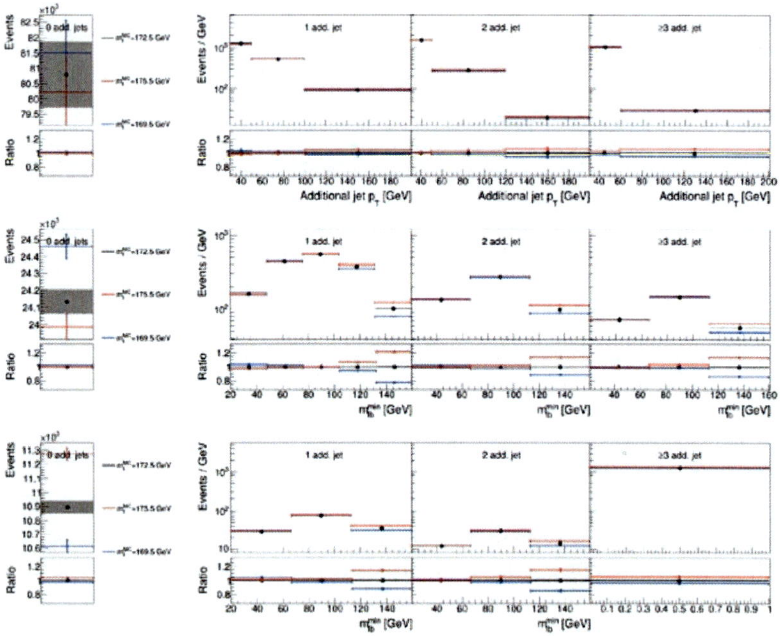

Fig. A.1 Pre-fit distributions for $m_t^{MC} = 172.5 \pm 3\,\text{GeV}$

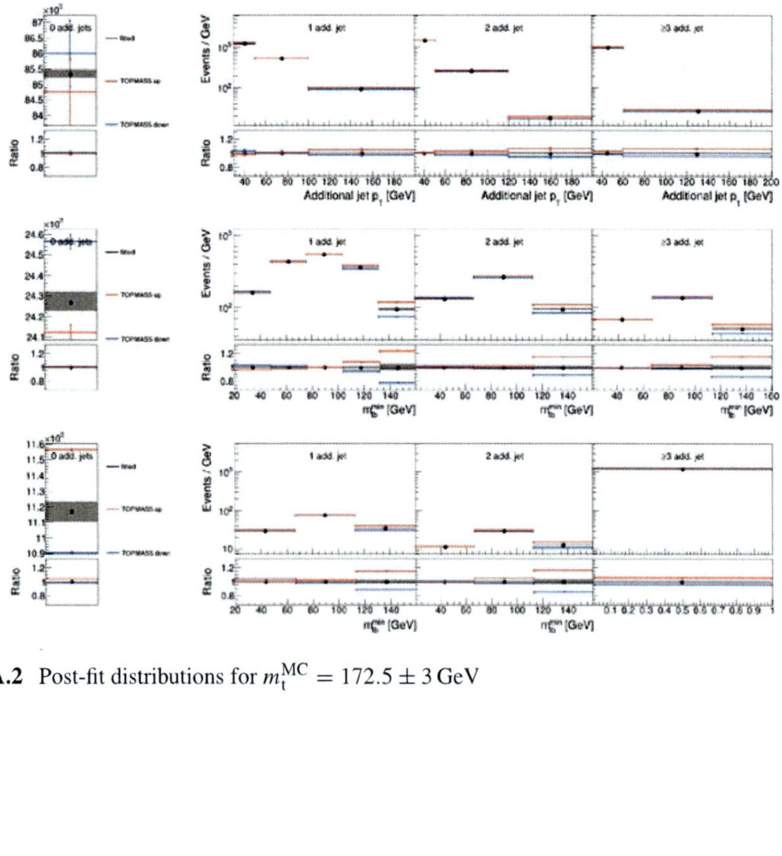

Fig. A.2 Post-fit distributions for $m_t^{MC} = 172.5 \pm 3\,\text{GeV}$

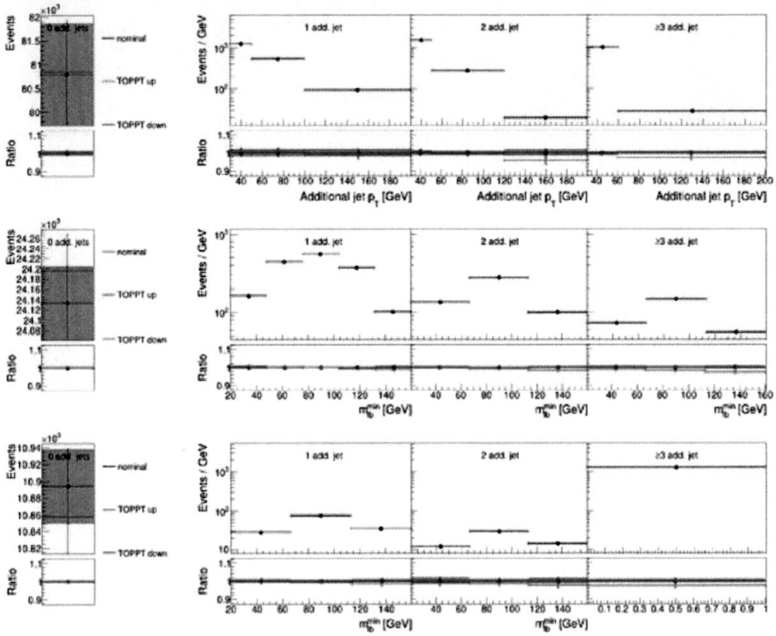

Fig. A.3 Pre-fit distributions for top quark p_T reweighting

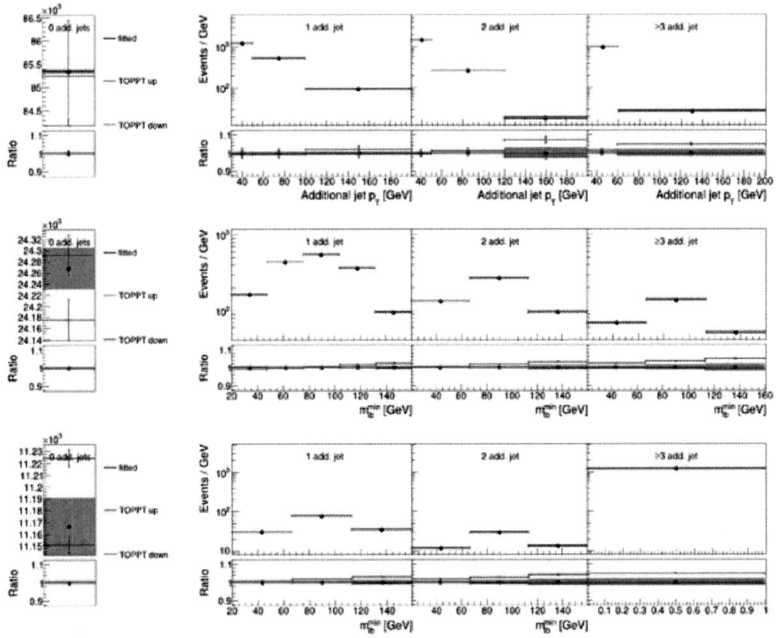

Fig. A.4 Post-fit distributions for top quark p_T reweighting

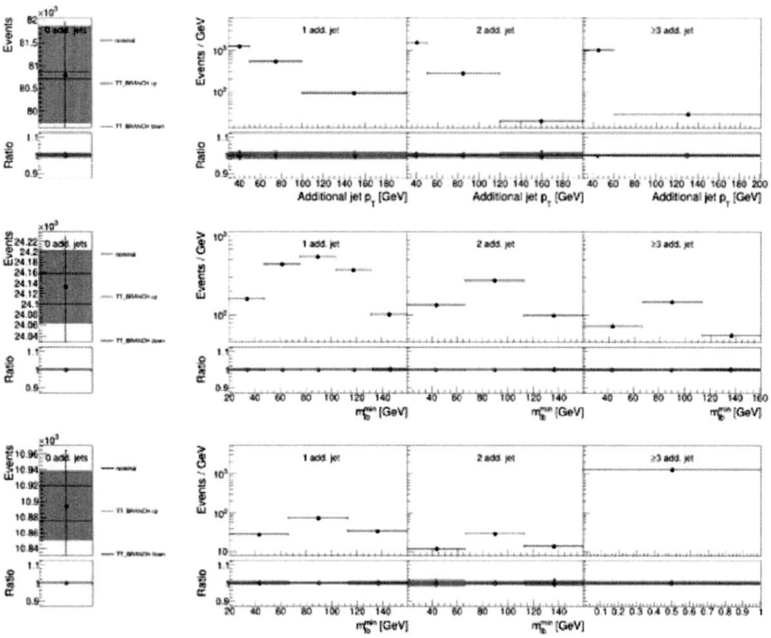

Fig. A.5 Pre-fit distributions for b-hadron branching fraction

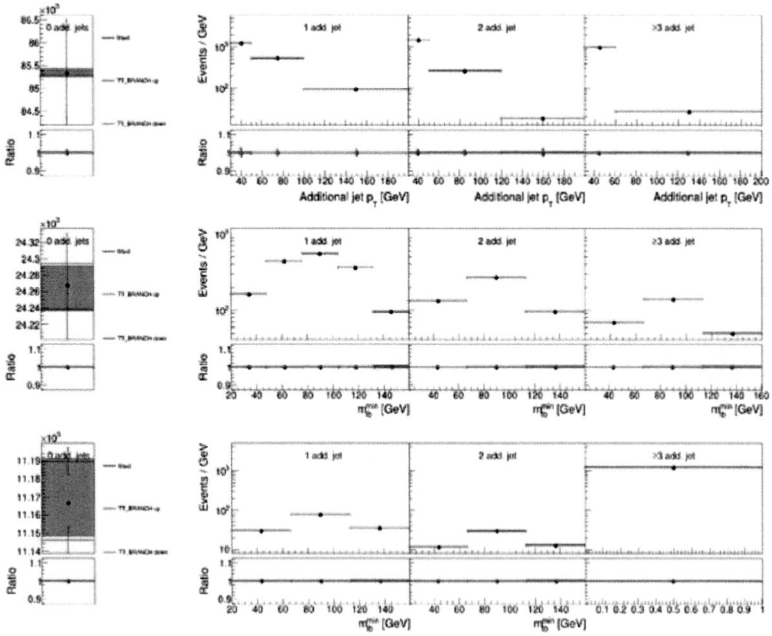

Fig. A.6 Post-fit distributions for b-hadron branching fraction

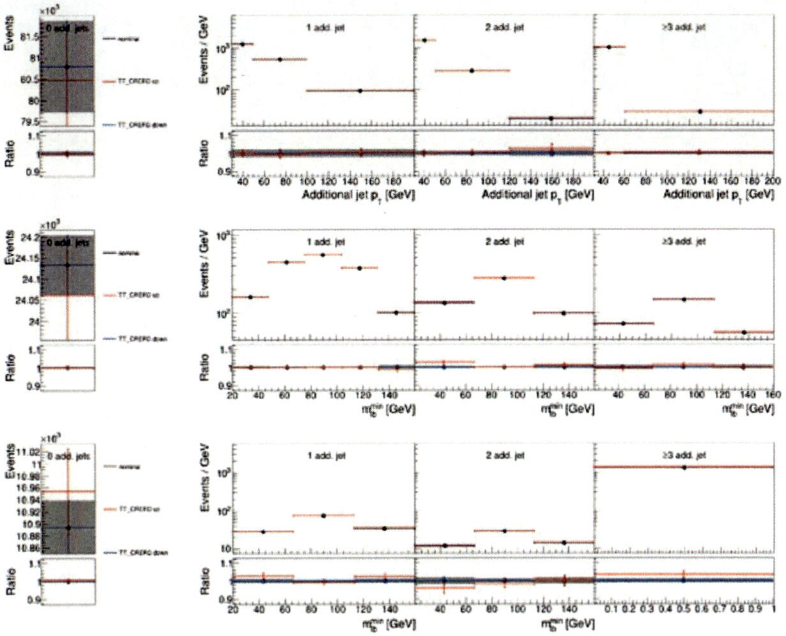

Fig. A.7 Pre-fit distributions for the CR model with ERD

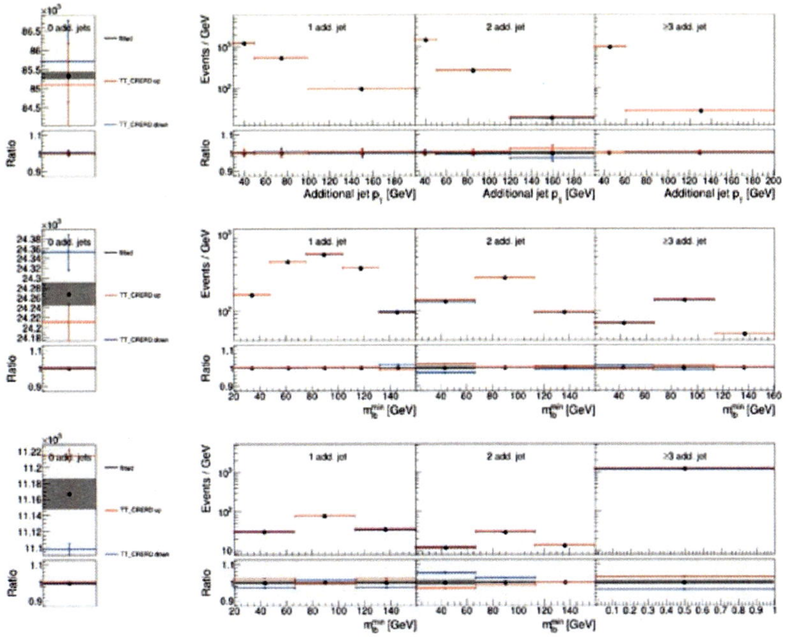

Fig. A.8 Post-fit distributions for the CR model with ERD

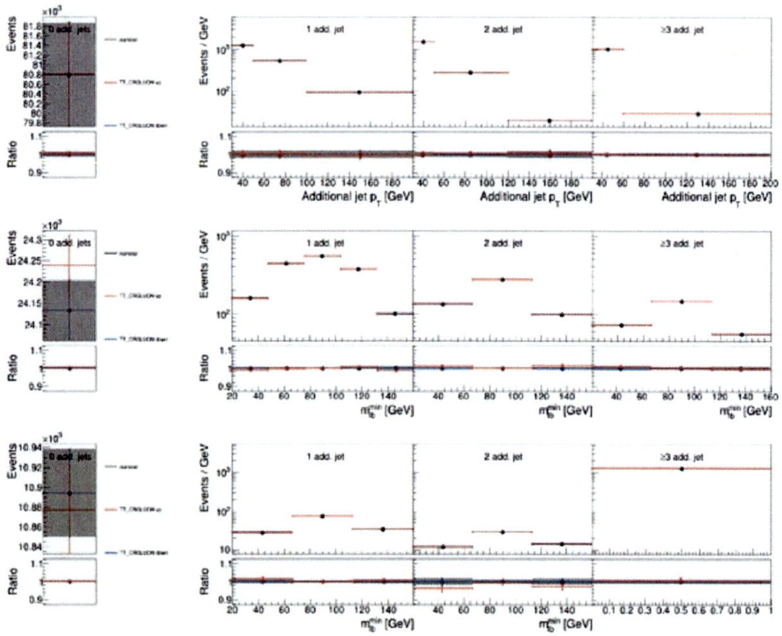

Fig. A.9 Pre-fit distributions for the gluon move CR model

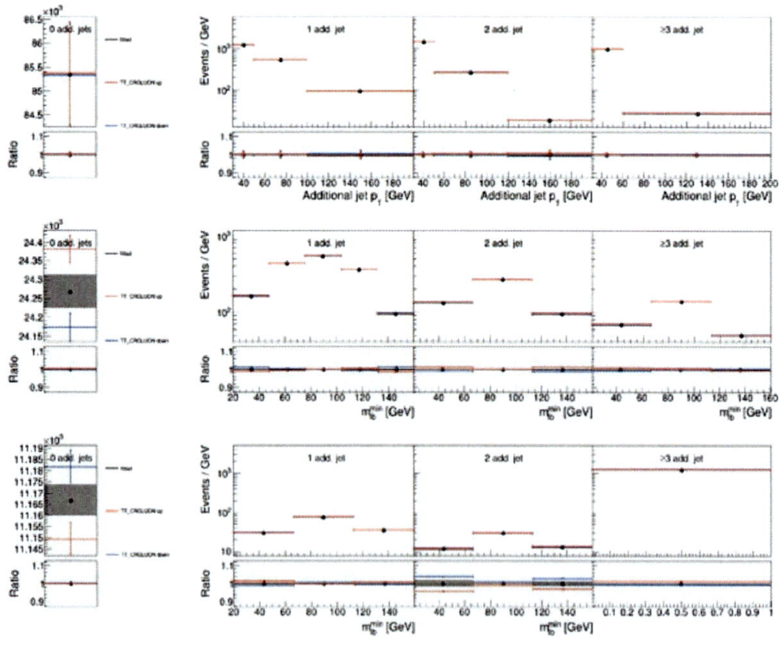

Fig. A.10 Post-fit distributions for the gluon move CR model

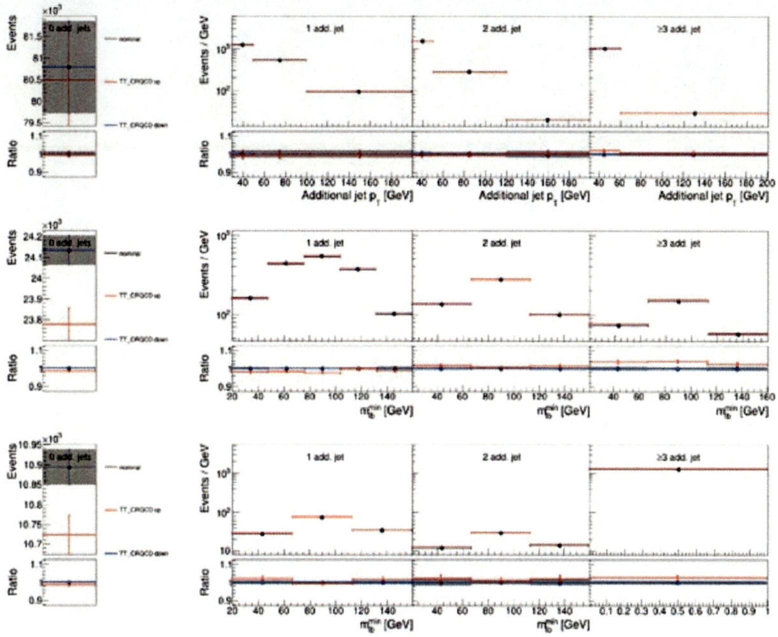

Fig. A.11 Pre-fit distributions for the QCD-inspired CR model

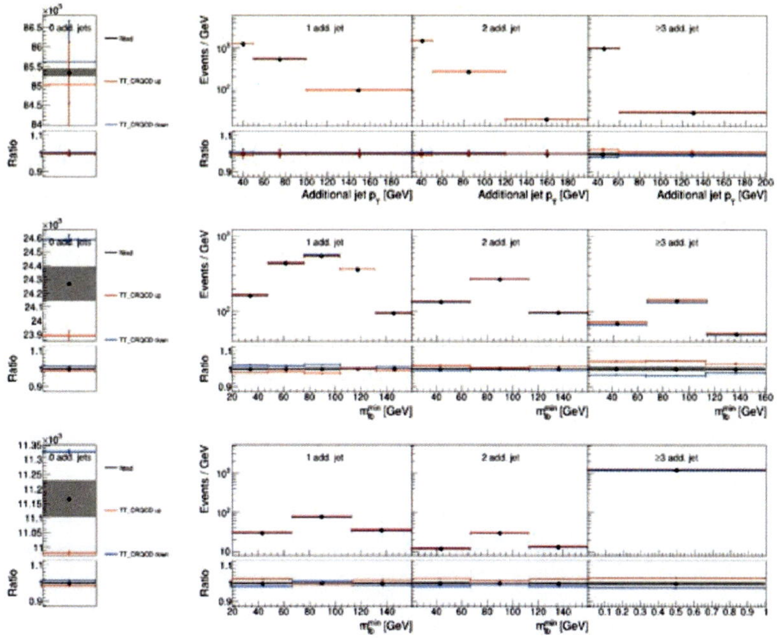

Fig. A.12 Post-fit distributions for the QCD-inspired CR model

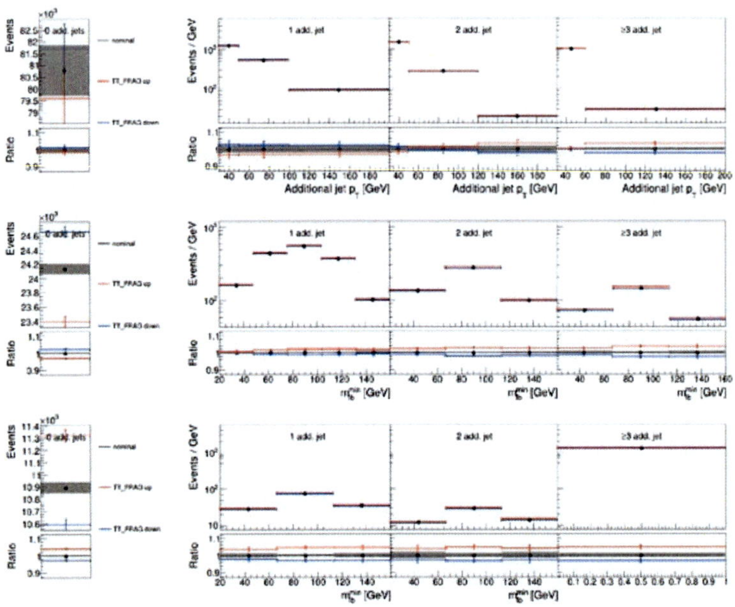

Fig. A.13 Pre-fit distributions for Bowler–Lund fragmentation function

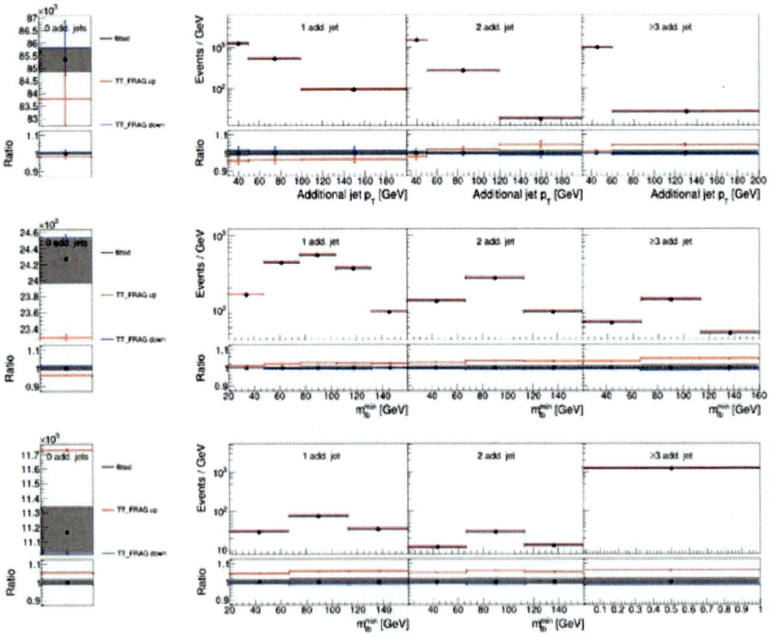

Fig. A.14 Post-fit distributions for Bowler–Lund fragmentation function

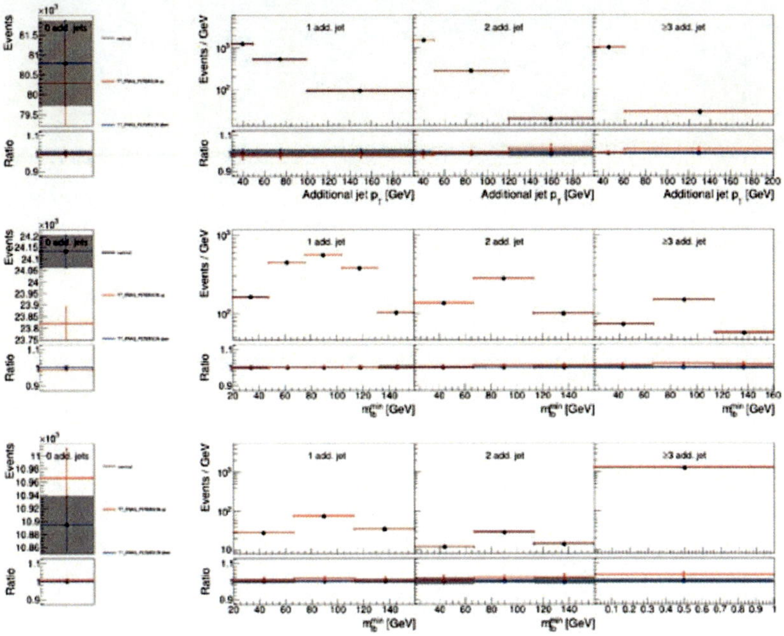

Fig. A.15 Pre-fit distributions for Peterson fragmentation function

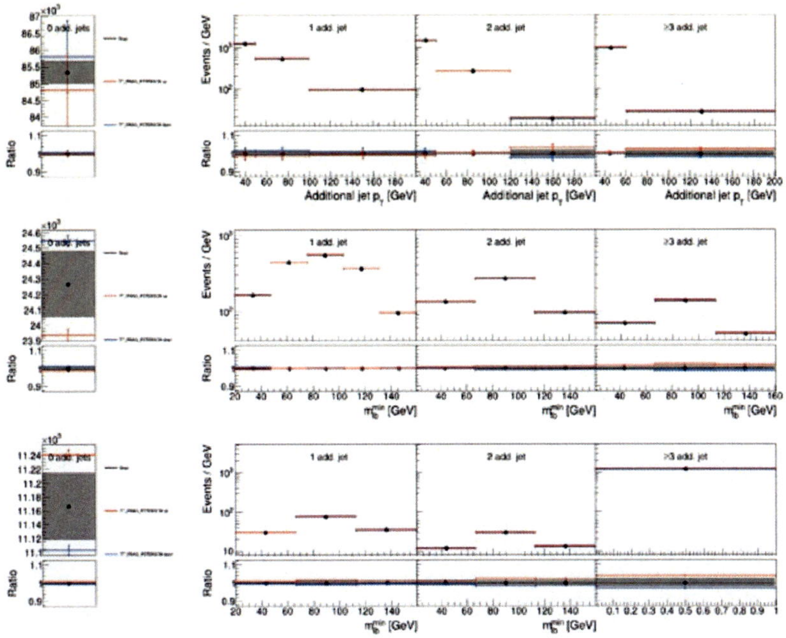

Fig. A.16 Post-fit distributions for Peterson fragmentation function

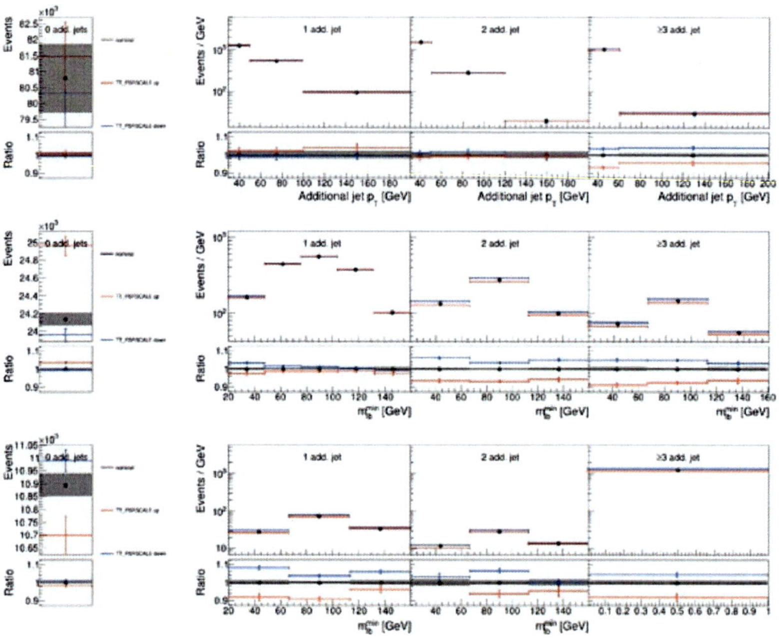

Fig. A.17 Pre-fit distributions for $t\bar{t}$ FSR scale

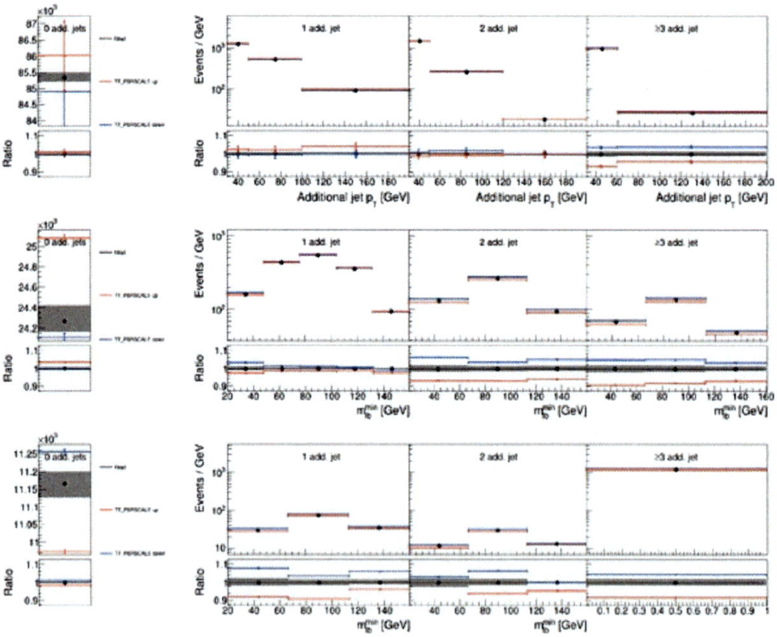

Fig. A.18 Post-fit distributions for $t\bar{t}$ FSR scale

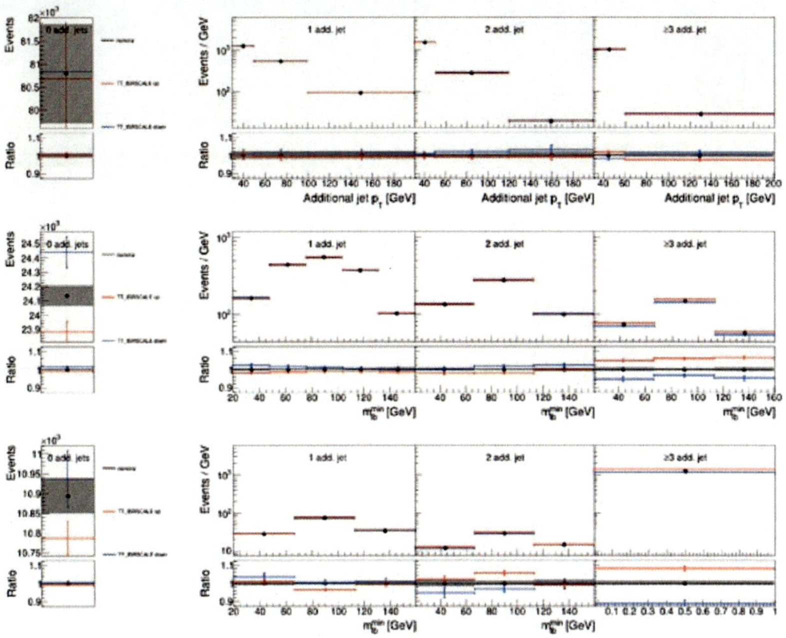

Fig. A.19 Pre-fit distributions for $t\bar{t}$ ISR scale

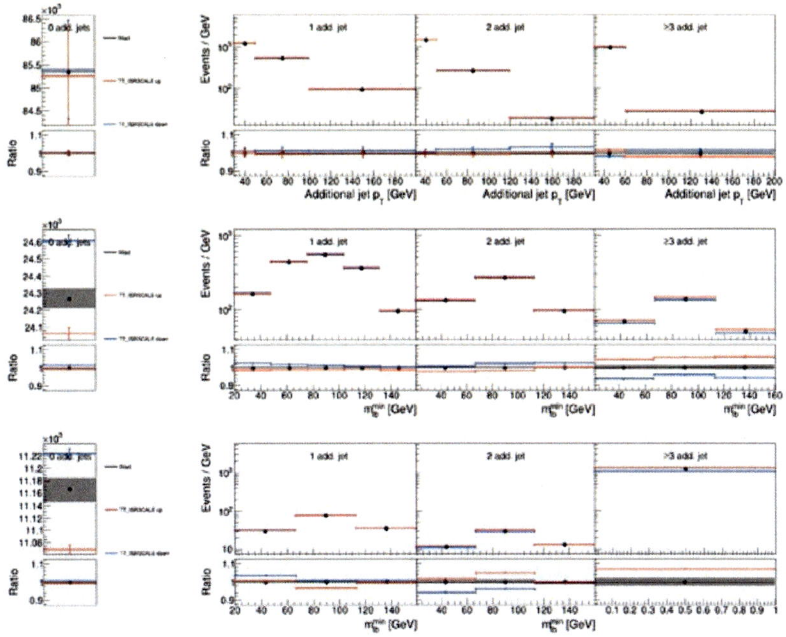

Fig. A.20 Post-fit distributions for $t\bar{t}$ ISR scale

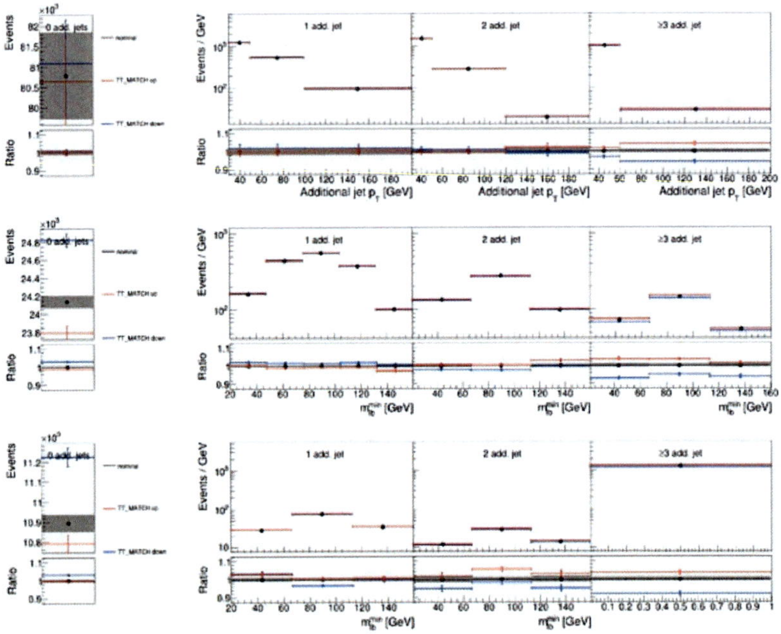

Fig. A.21 Pre-fit distributions for ME/PS matching

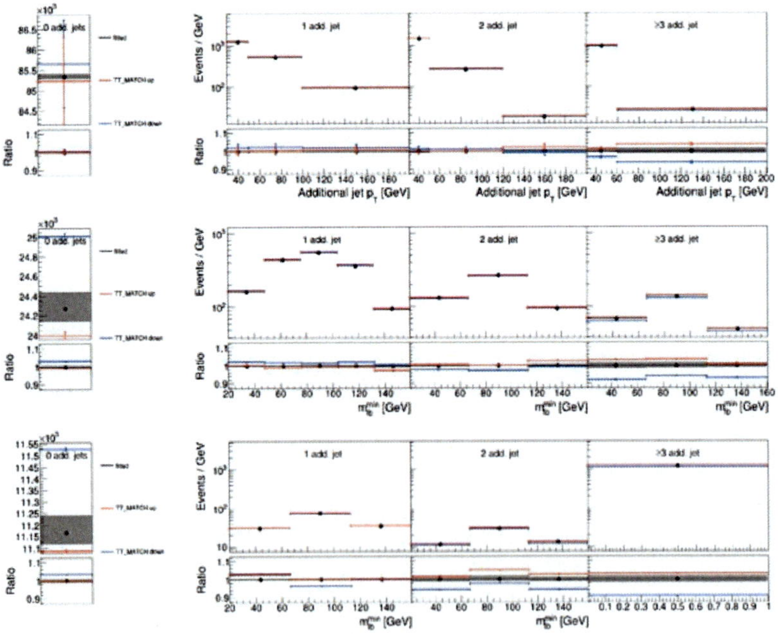

Fig. A.22 Post-fit distributions for ME/PS matching

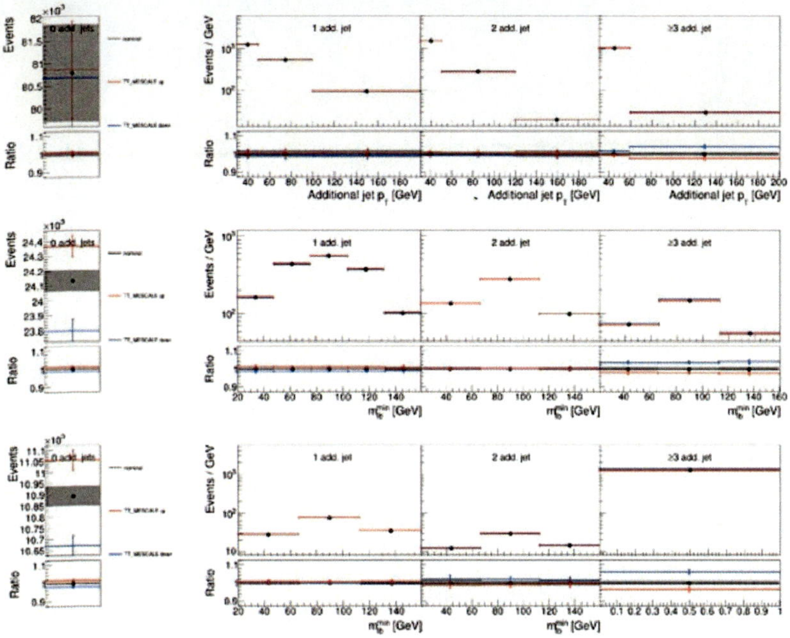

Fig. A.23 Pre-fit distributions for $t\bar{t}$ ME scale

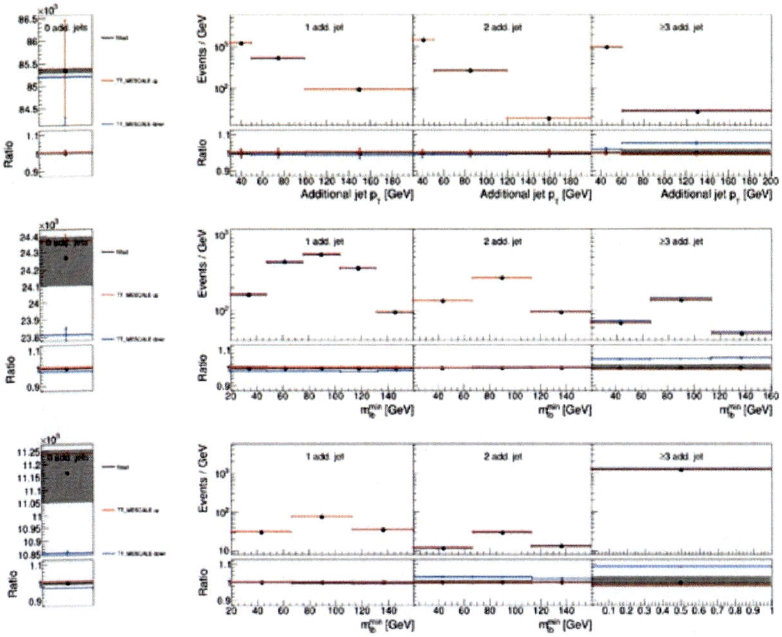

Fig. A.24 Post-fit distributions for $t\bar{t}$ ME scale

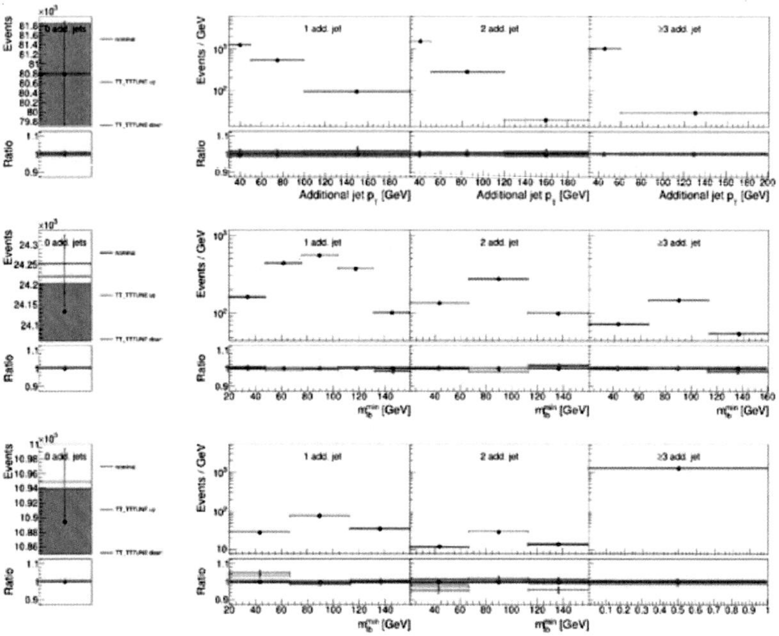

Fig. A.25 Pre-fit distributions for UE tune

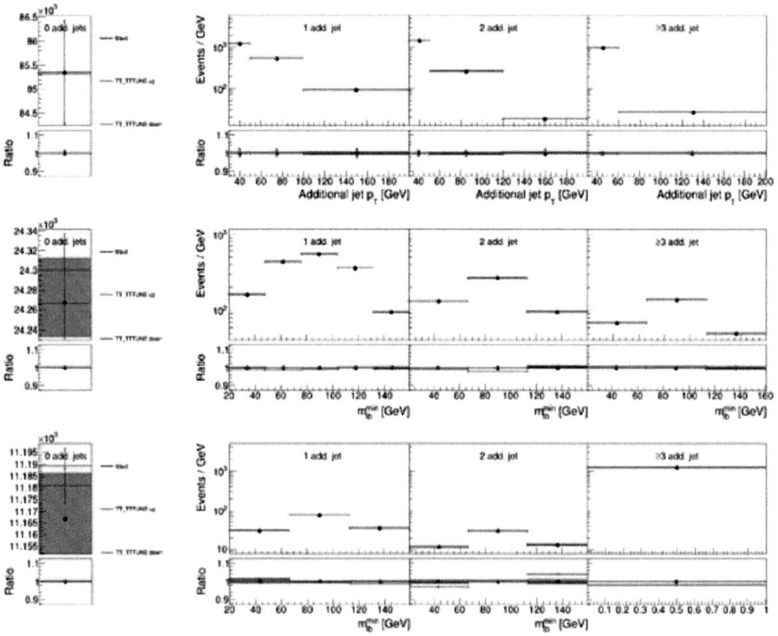

Fig. A.26 Post-fit distributions for UE tune

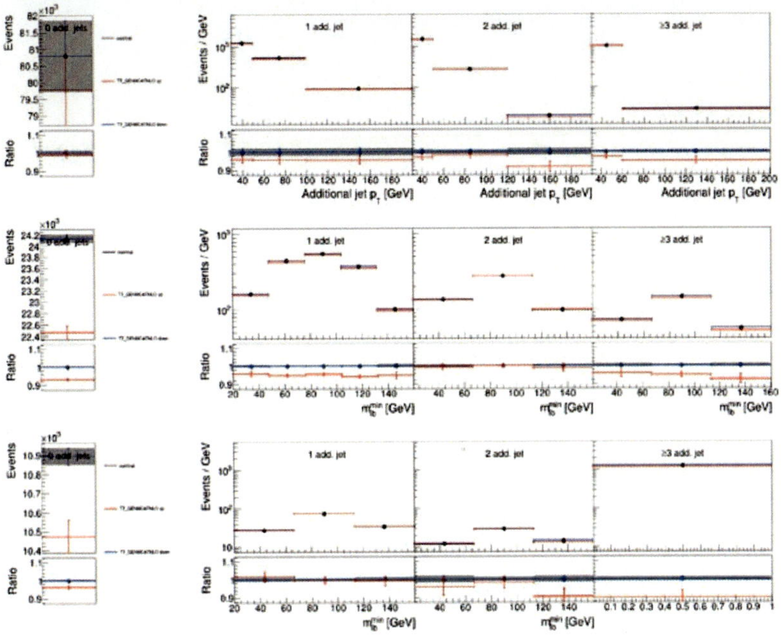

Fig. A.27 Pre-fit distributions for MG5_AMC@NLO generator

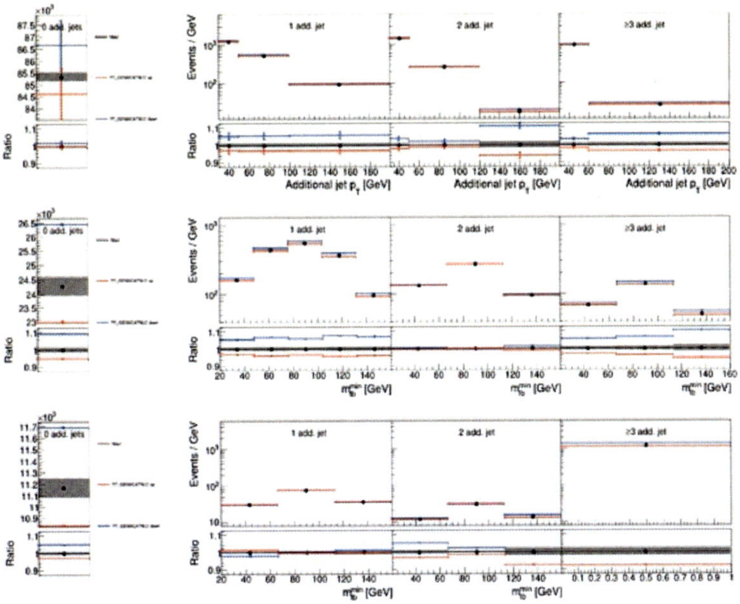

Fig. A.28 Post-fit distributions for MG5_AMC@NLO generator

Appendix B
Closure Test of the Simultaneous Fit of $\sigma_{t\bar{t}}$ and m_t^{MC}

The closure of the fit described in Chap. 5 is tested by performing fits where data are replaced by the sum of the MC templates for the signal and the backgrounds. The results obtained by replacing data with the nominal simulation can be interpreted as the expected results of the fit, and are shown in Fig. B.1. The details of the pull plots are explained in Chap. 5. The pulls on all nuisance parameters are equal or very close to zero, which indicates the good closure of the fit.

In all subsequent figures, the data are replaced by the sum of the nominal background templates and the $t\bar{t}$ templates corresponding to one of the modelling variations. In this case, the closure test is successful if the pull on the nuisance parameter related to the considered variation is equal or close to $+1$ (1) for up (down) and one-sided variations, while the pulls on all other nuisance parameters are equal or close to zero. Good closure is observed for all nuisance parameters, although the strong correlation between the Bowler–Lund and the Peterson models can be observed in the case of the b quark fragmentation.

M. M. Defranchis, *First Measurement of the Running of the Top Quark Mass*, Springer Theses, https://doi.org/10.1007/978-3-030-90376-3

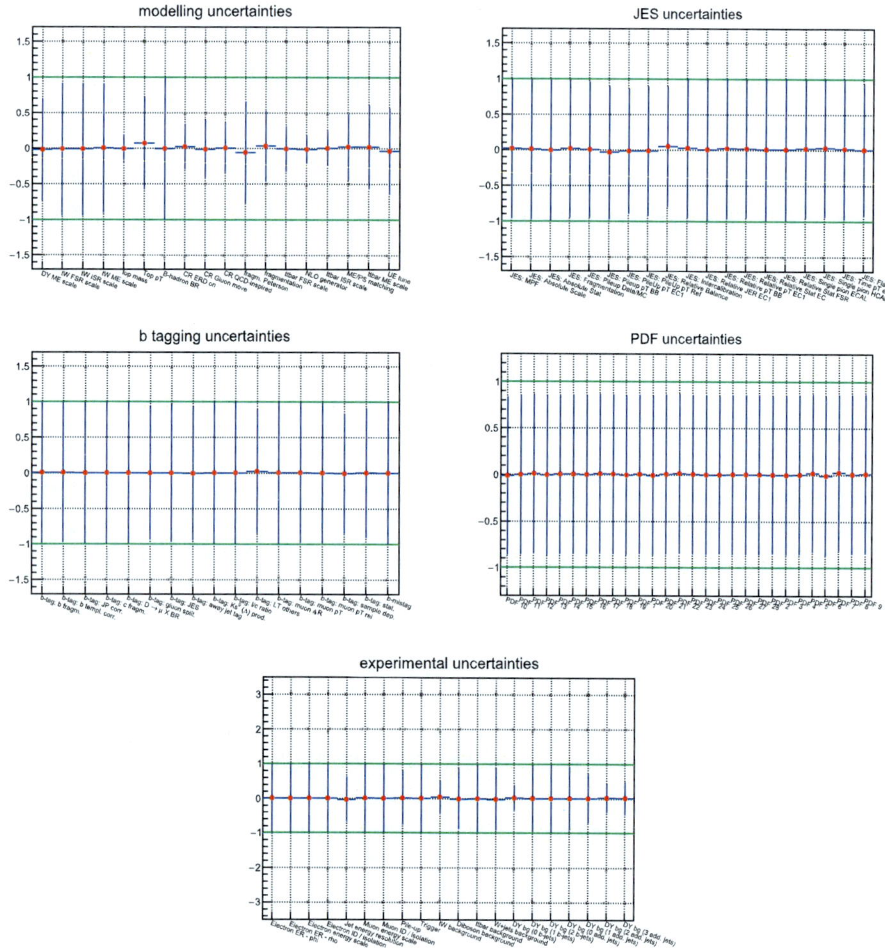

Fig. B.1 Fit closure test and expected constraints on the nuisance parameters related to modelling (upper left), JES (upper right), b tagging (middle left), PDF (middle right), and other experimental uncertainties (lower). The constraints on the PDFs uncertainties are due to the rescaling of corresponding variations from a 95 to a 68% confidence interval

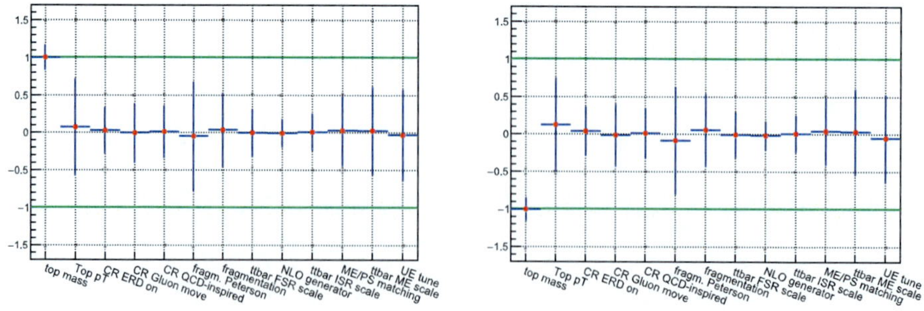

Fig. B.2 Closure test for $m_{\rm t}^{\rm MC}$ + 3 GeV (left) and $m_{\rm t}^{\rm MC}$ 3 GeV (right)

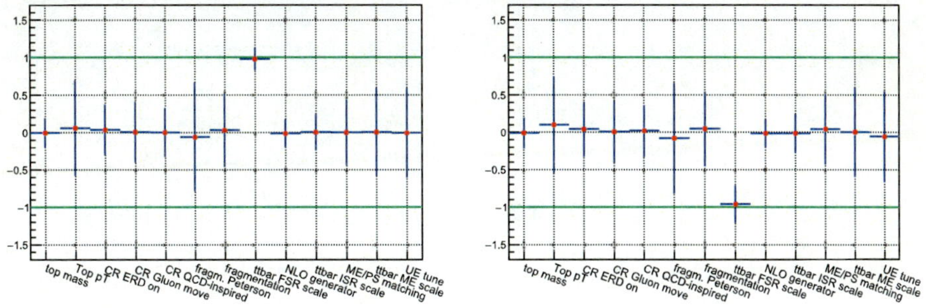

Fig. B.3 Closure test for FSR scale up (left) and down (right)

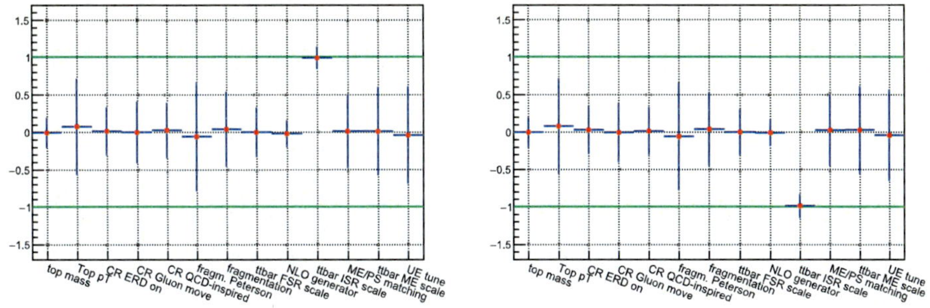

Fig. B.4 Closure test for ISR scale up (left) and down (right)

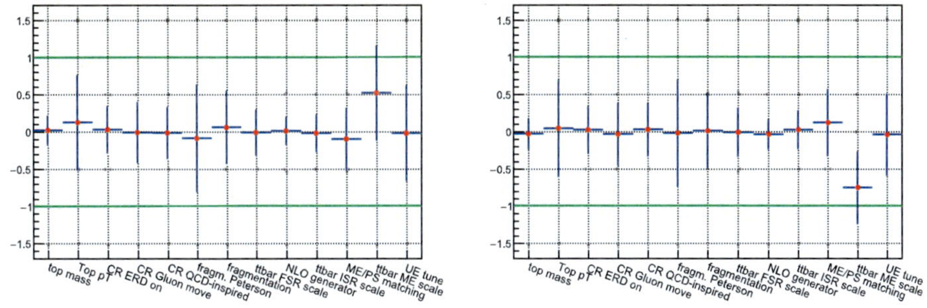

Fig. B.5 Closure test for ME scale up (left) and down (right)

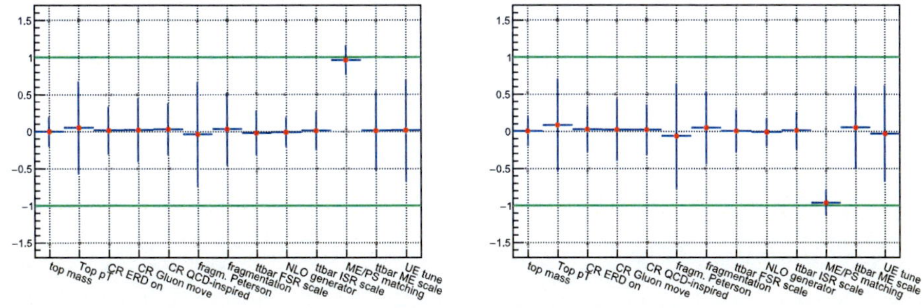

Fig. B.6 Closure test for ME/PS matching up (left) and down (right)

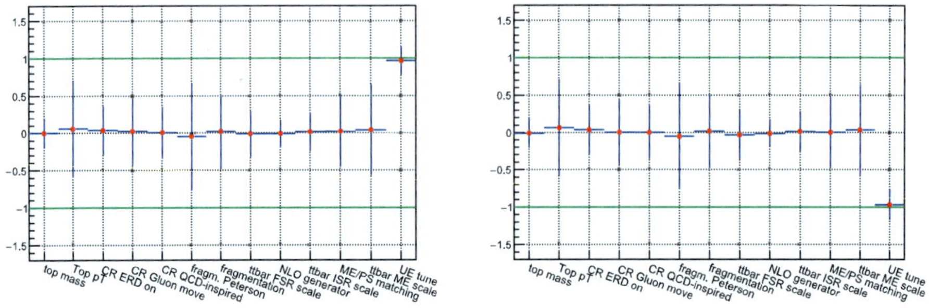

Fig. B.7 Closure test for UE tune up (left) and down (right)

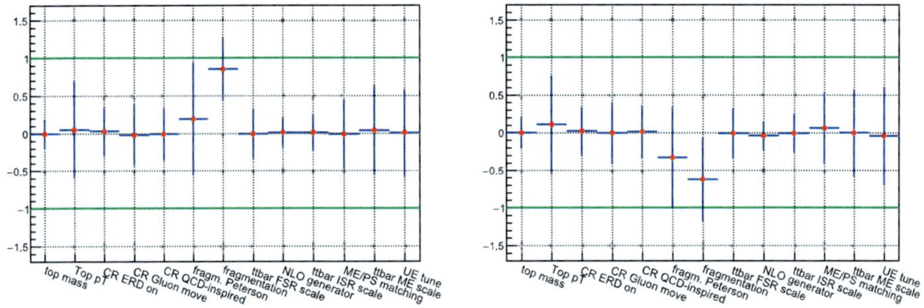

Fig. B.8 Closure test b quark fragmentation up (left) and down (right)

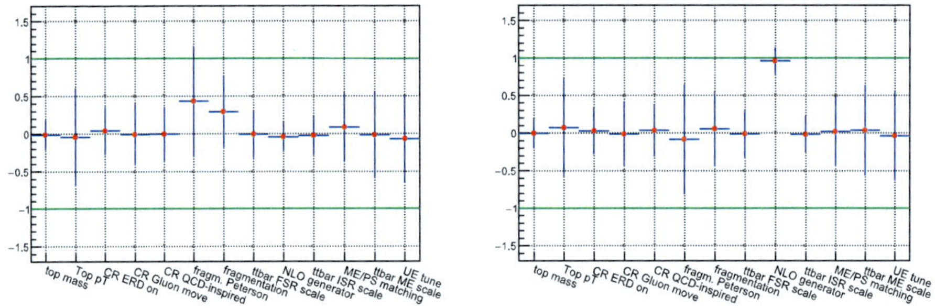

Fig. B.9 Closure test for Peterson fragmentation (left) and MG5_aMC@NLO generator (right)

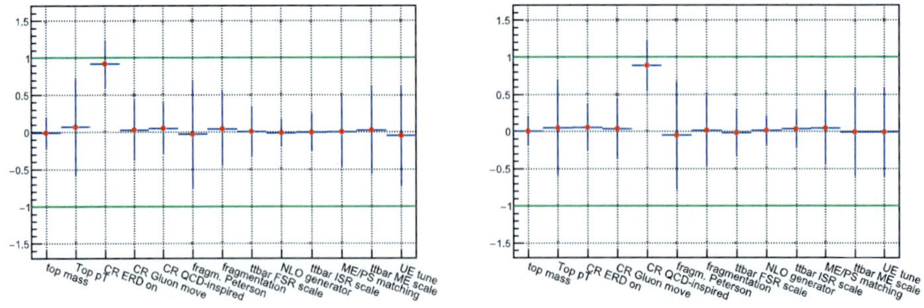

Fig. B.10 Closure test for the CR model with ERD (left) and QCD-inspired CR model (right)

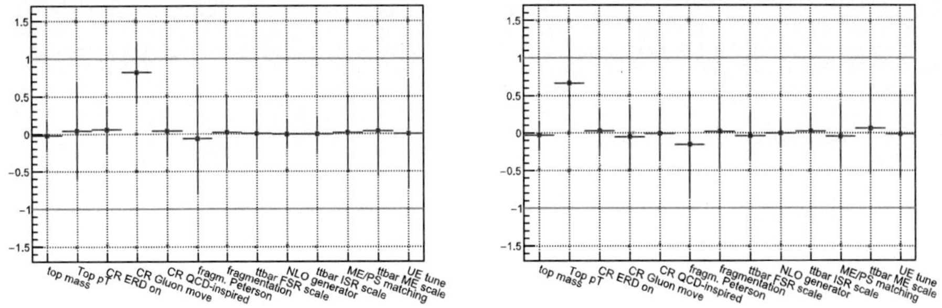

Fig. B.11 Closure test for the gluon move CR model (left) and top p_T reweighting (right)

Appendix C
Cross-Check Fit of m_t^{MC} with a Single Distribution

In this appendix, additional details on the result of the cross check fit of m_t^{MC} described in Chap. 5 are provided. In particular, the pulls and constraints on the nuisance parameters are shown in Fig. C.1 (with the exception of those related to the modelling uncertainties, which are given in Chap. 5), and the impact of the various sources of systematic uncertainty on the measured m_t^{MC} is summarized in Table C.1. The details of the plots and table can be found in Chap. 5.

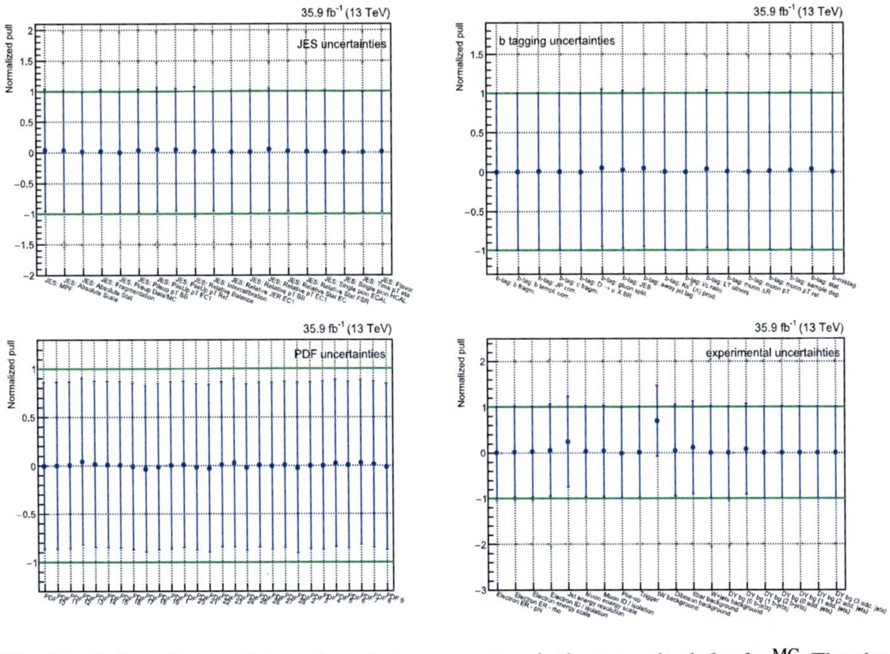

Fig. C.1 Pulls and constraints on the nuisance parameters in the cross-check fit of m_t^{MC}. The plots refer to the JES (upper left), b tagging (upper right), PDF (lower left), and other experimental uncertainties (lower right)

M. M. Defranchis, *First Measurement of the Running of the Top Quark Mass*, Springer Theses, https://doi.org/10.1007/978-3-030-90376-3

Table C.1 The absolute uncertainties in m_t^{MC} and their sources, as determined in the cross-check fit of m_t^{MC}

Source	Uncertainty (GeV)
Trigger	0.01
Lepton ident./isolation	0.01
Muon momentum scale	0.04
Electron momentum scale	0.05
Jet energy scale	0.67
Jet energy resolution	0.16
b tagging	0.05
Pileup	0.01
$t\bar{t}$ ME scale	0.01
tW ME scale	0.01
DY ME scale	0.01
NLO generator	0.07
PDF	0.06
$\sigma_{t\bar{t}}$	0.07
Top quark p_T	0.17
ME/PS matching	0.19
UE tune	0.08
$t\bar{t}$ ISR scale	0.09
tW ISR scale	0.01
$t\bar{t}$ FSR scale	0.06
tW FSR scale	0.01
b quark fragmentation	0.05
b hadron BF	0.05
Colour reconnection	0.08
DY background	0.02
tW background	0.02
Diboson background	0.01
$t\bar{t}$ background	0.01
W+jets background	0.01
Statistical	0.13
MC statistical	0.14
Total m_t^{MC} uncertainty	$\pm^{0.76}_{0.77}$

Appendix D
Impact of MC Statistical Fluctuations on the Pulls and Constraints of the Modelling Uncertainties

In this appendix, the effect of the MC statistical uncertainty on the pulls and constraints on the main $t\bar{t}$ modelling uncertainties is investigated. The distributions are derived using the method described in Chap. 5. The results show that the constraints can be significantly affected by the uncertainty in the template morphing. As in the case of $\sigma_{t\bar{t}}$ and m_t^{MC}, the pull distributions exhibit a quasi-Gaussian shape centred at the best-fit value of the nuisance parameters. A small secondary structure is observed in the case of the UE tune, which is estimated using a sample with very large statistical uncertainty. This is due to the fact that, in the presence of large statistical fluctuations, different local minima can be found in the fit.

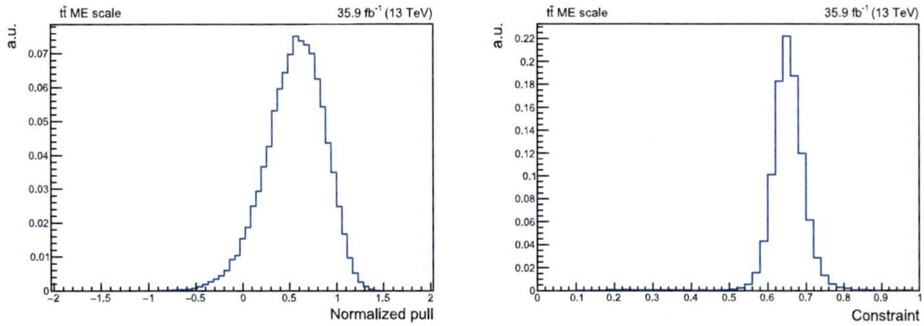

Fig. D.1 Distributions of the pulls (left) and constraints (right) on the $t\bar{t}$ ME scale uncertainty obtained using toy templates

M. M. Defranchis, *First Measurement of the Running of the Top Quark Mass*, Springer Theses, https://doi.org/10.1007/978-3-030-90376-3

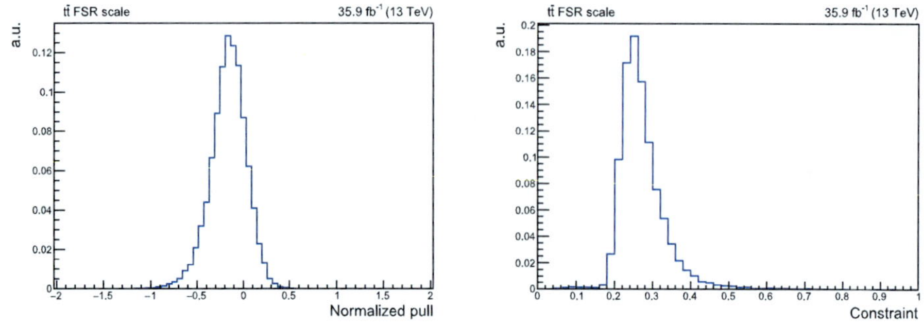

Fig. D.2 Distributions of the pulls (left) and constraints (right) on the $t\bar{t}$ FSR scale uncertainty obtained using toy templates

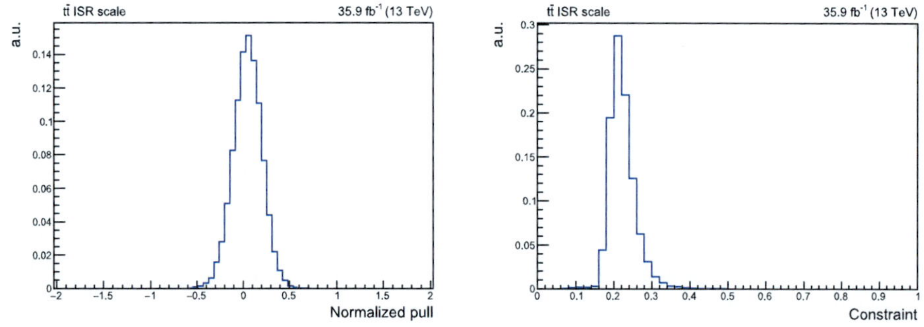

Fig. D.3 Distributions of the pulls (left) and constraints (right) on the $t\bar{t}$ ISR scale uncertainty obtained using toy templates

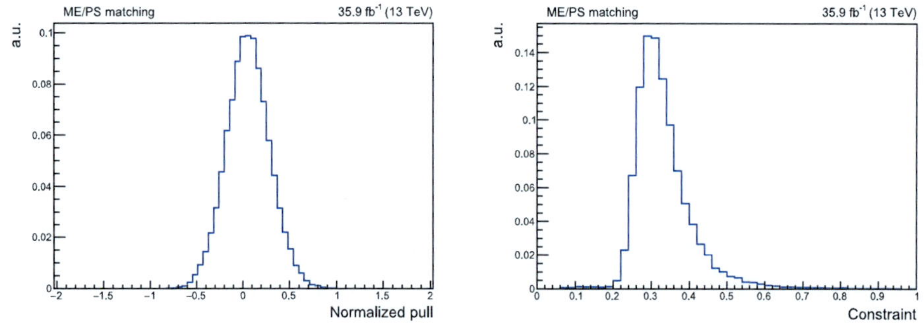

Fig. D.4 Distributions of the pulls (left) and constraints (right) on the ME/PS matching uncertainty obtained using toy templates

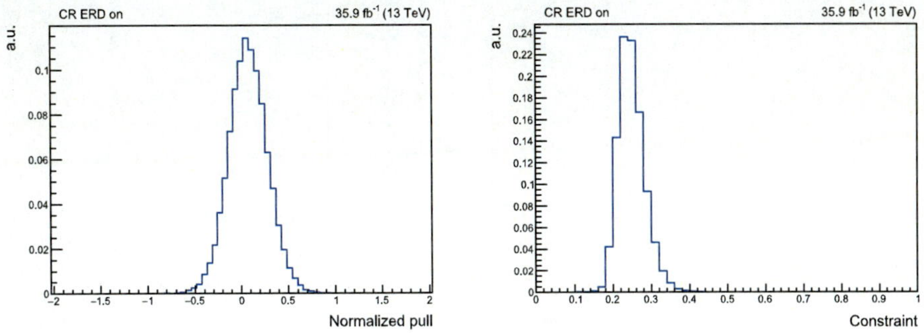

Fig. D.5 Distributions of the pulls (left) and constraints (right) on the CR model with ERD obtained using toy templates

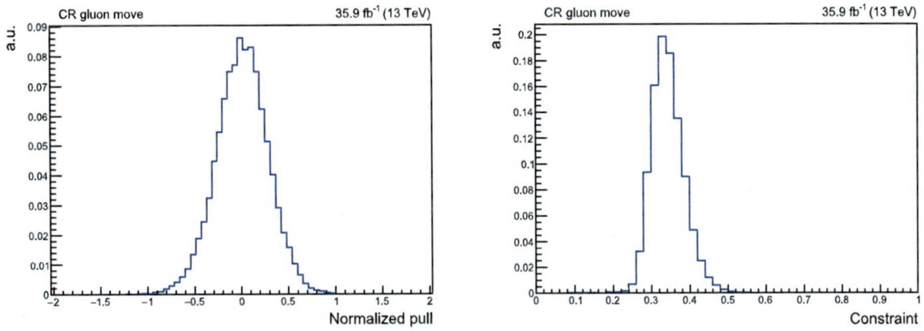

Fig. D.6 Distributions of the pulls (left) and constraints (right) on the gluon move CR model obtained using toy templates

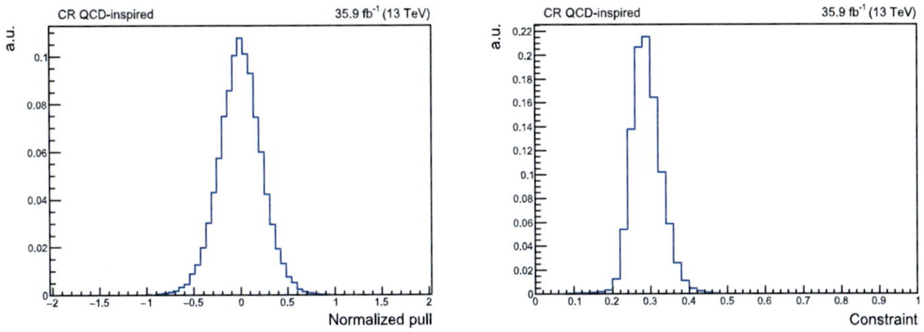

Fig. D.7 Distributions of the pulls (left) and constraints (right) on the CR (QCD-inspired) uncertainty obtained using toy templates

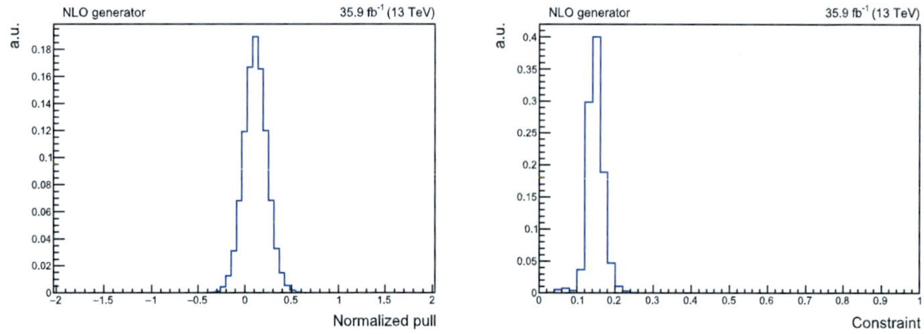

Fig. D.8 Distributions of the pulls (left) and constraints (right) on the NLO generator uncertainty obtained using toy templates

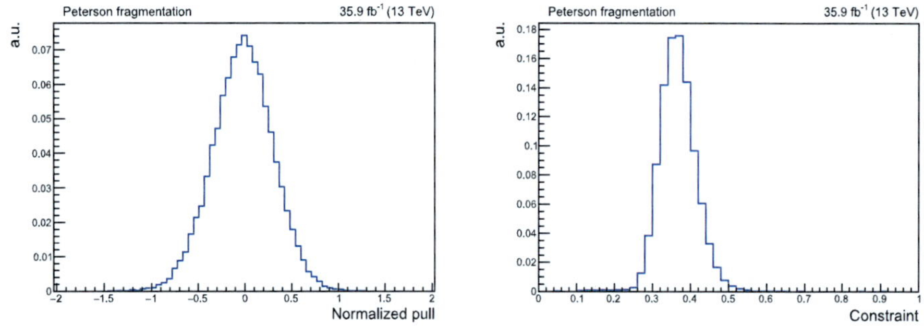

Fig. D.9 Distributions of the pulls (left) and constraints (right) on the Peterson fragmentation uncertainty obtained using toy templates

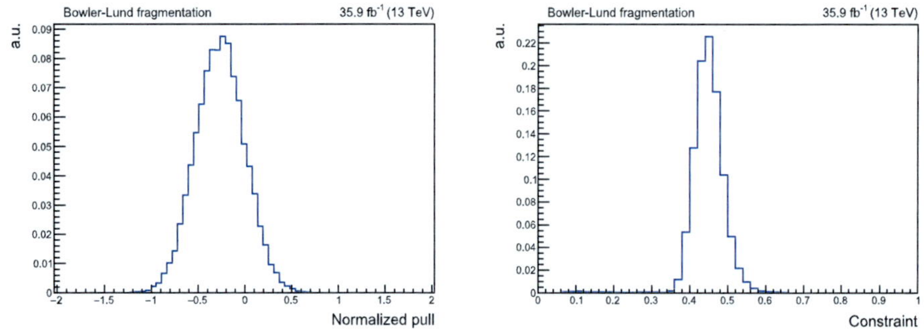

Fig. D.10 Distributions of the pulls (left) and constraints (right) on the Bowler–Lund fragmentation uncertainty obtained using toy templates

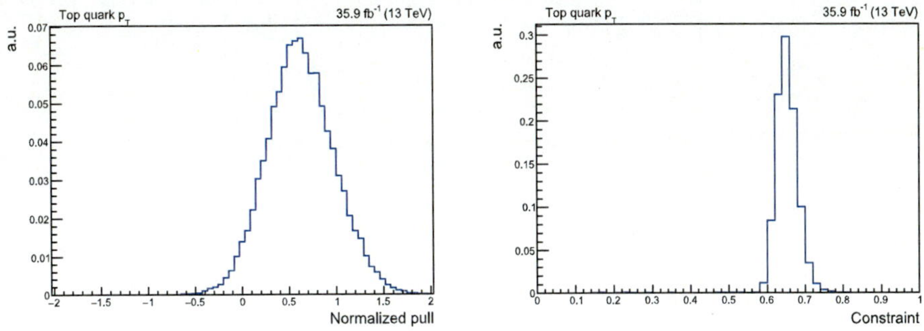

Fig. D.11 Distributions of the pulls (left) and constraints (right) on the top quark p_T uncertainty obtained using toy templates

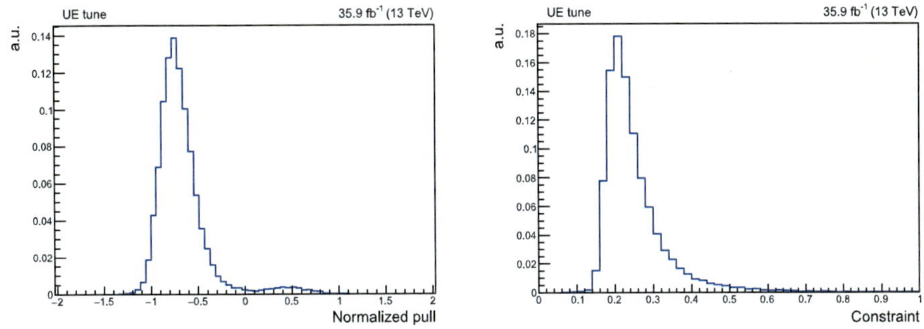

Fig. D.12 Distributions of the pulls (left) and constraints (right) on the UE tune uncertainty obtained using toy templates

Appendix E
Cross-Check Measurement of the Differential $\mathrm{t\bar{t}}$ Cross Section with the Loose Kinematic Reconstruction

The measurement of $\mathrm{d}\sigma_{\mathrm{t\bar{t}}}/\mathrm{d}m_{\mathrm{t\bar{t}}}$ presented in Chap. 7 is cross-checked by repeating the analysis using an alternative kinematic reconstruction algorithm which does not depend on the parameter $m_{\mathrm{t}}^{\mathrm{kin}}$. In this algorithm, described in Ref. [1], the momentum of the $\nu\bar{\nu}$ system is reconstructed rather than the momenta of each neutrino. This information is sufficient to estimate the quantity $m_{\mathrm{t\bar{t}}}^{\mathrm{reco}}$ and all other variables related to the kinematics of the $\mathrm{t\bar{t}}$ system as a whole.

The algorithm examines all possible combinations of leptons and jets with an invariant mass lower than 180 GeV, which are then ranked according to the number of b-tagged jets. If different combinations with the same number of b-tagged jets are found, those with the jets of highest $\mathrm{p_T}$ are chosen. The momentum of the $\nu\bar{\nu}$ system is then estimated by assuming that its transverse component corresponds to the missing transverse energy in the event and its longitudinal one to that of the lepton pair. In addition, the invariant mass of the $\nu\bar{\nu}$ pair is required to be positive, and that of the reconstructed $\mathrm{W^{+}W}$ system to be larger than twice the mass of the W boson. The resolution and efficiency of this algorithm are found to be similar to the one described in Chap. 7 [1].

The distribution of $m_{\mathrm{t\bar{t}}}^{\mathrm{reco}}$ after the fit to the data is shown in Fig. E.1. The result for $\mathrm{d}\sigma_{\mathrm{t\bar{t}}}/\mathrm{d}m_{\mathrm{t\bar{t}}}$ can be found in Fig. E.2, and the values of the $\sigma_{\mathrm{t\bar{t}}}^{(\mu_k)}$ are measured to be:

$$
\begin{aligned}
\sigma_{\mathrm{t\bar{t}}}^{(\mu_1)} &= 257 \pm 13\,(\mathrm{syst}) \pm 2\,(\mathrm{stat})\ \mathrm{pb},\\
\sigma_{\mathrm{t\bar{t}}}^{(\mu_2)} &= 308 \pm 17\,(\mathrm{syst}) \pm 2\,(\mathrm{stat})\ \mathrm{pb},\\
\sigma_{\mathrm{t\bar{t}}}^{(\mu_3)} &= 190 \pm 10\,(\mathrm{syst}) \pm 2\,(\mathrm{stat})\ \mathrm{pb},\\
\sigma_{\mathrm{t\bar{t}}}^{(\mu_4)} &= 50 \pm 4\,(\mathrm{syst}) \pm 1\,(\mathrm{stat})\ \mathrm{pb}.
\end{aligned}
$$

M. M. Defranchis, *First Measurement of the Running of the Top Quark Mass*, Springer Theses, https://doi.org/10.1007/978-3-030-90376-3

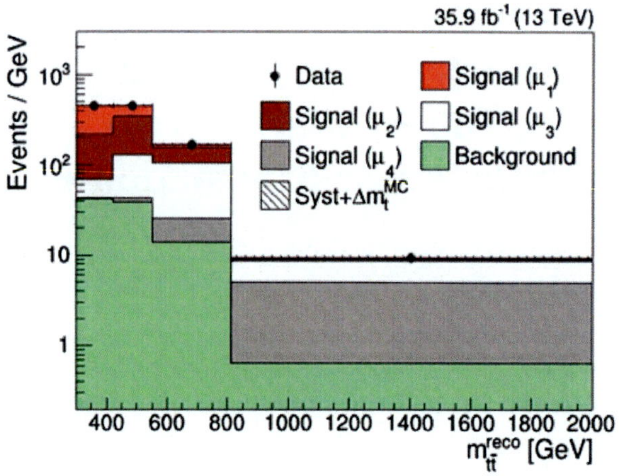

Fig. E.1 The same as Fig. 7.1, but for the cross-check measurement performed with the loose kinematic reconstruction

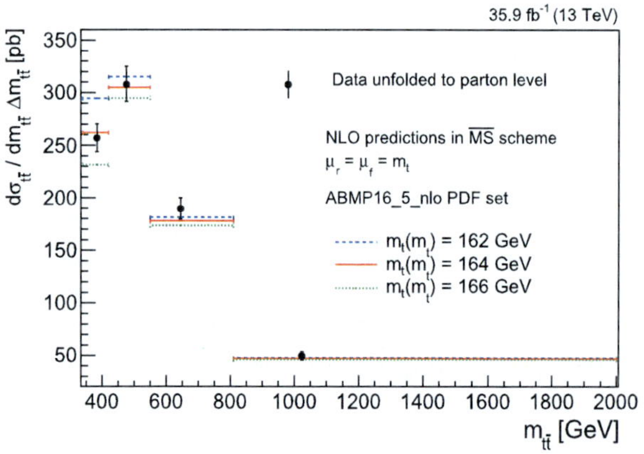

Fig. E.2 The same as Fig. 7.6, but for the cross-check measurement performed with the loose kinematic reconstruction

The results are found to be in agreement with those reported in Chap. 7, although with slightly larger systematic uncertainties. For this reason, the full kinematic reconstruction of Chap. 7 is used in the main result.

Finally, the pulls and constraints on the modelling and other nuisance parameters are shown in Figs. E.3 and E.4, respectively. Most pulls and constraints are found to be compatible with those obtained in Chap. 7.

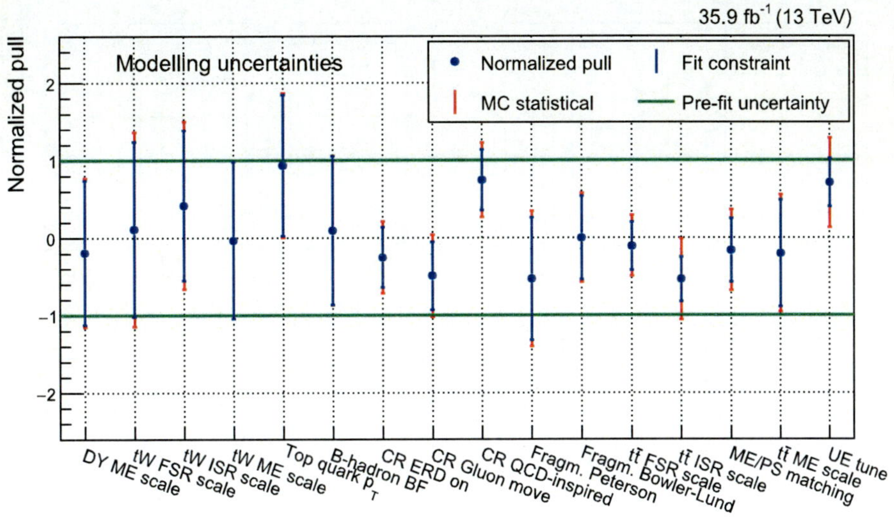

Fig. E.3 The same as Fig. 7.4, but for the cross-check measurement performed with the loose kinematic reconstruction algorithm

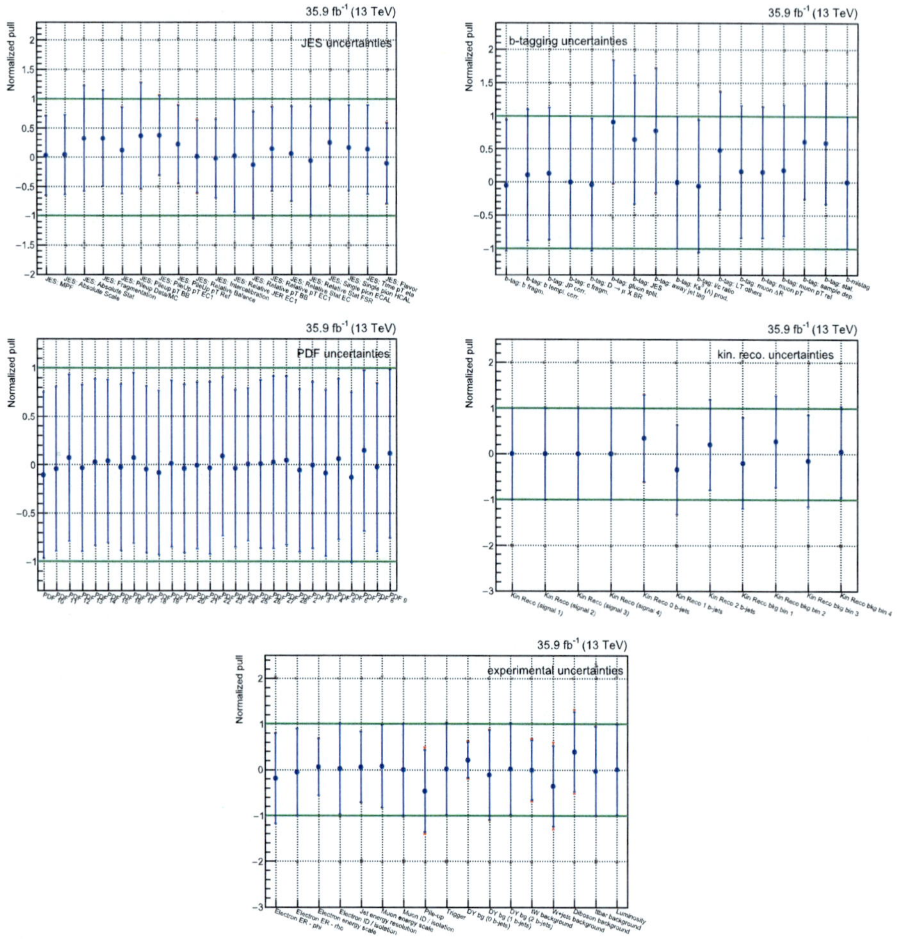

Fig. E.4 The same as Fig. 7.5, but for the cross-check measurement performed with the loose kinematic reconstruction algorithm

Reference

1. CMS Collaboration (2019) Measurement of $t\bar{t}$ normalised multi-differential cross sections in pp collisions at $\sqrt{s} = 13$ TeV, and simultaneous determination of the strong coupling strength, top quark pole mass, and parton distribution functions. Eur. Phys. J. arXiv:1904.05237